MASTERING

WITH iOS 13

Updated Tips and Tricks to Operate Your

iPhone XR in iOS 13

Tech Reviewer

TABLE OF CONTENT

How to Use this Book

Welcome! Thank you for purchasing this book and trusting us to lead you right in operating your new device. This book has covered every details and tips you need to know about the iPhone XR for to get the best from the device.

To better understand how the book is structured, I would advise you read from page to page after which you can then navigate to particular sections as well as make reference to a topic individually. This book has been written in the simplest form to ensure that every user understands and gets the best out of this book. The table of content is also well outlined to make it easy for you to reference topics as needed at the speed of light.

Thank you.

Other Books by Same Author

- Fire TV Stick; 2019 Complete User Guide to Master the Fire Stick, Install Kodi and Over 100 Tips and Tricks https://amzn.to/2FnmcQ9
- Mastering Your iPhone X: iPhone X User Guide for Beginners and Seniors (2019 Version) https://amzn.to/2J1ywGW
- Amazon Echo Dot 3rd Generation: Advanced User Guide to Master Your Device with Instructions, Tips and Tricks https://amzn.to/31PaBTF

Introduction

In 2018, Apple followed the footsteps of iPhone XS and XS Max to launch the iPhone XR which was tagged the "Cheapest iPhone device of the year." This alone was enough to make users excited about owning a new iPhone XR. However, it is important to note that the term "cheap" does not mean the device has a low standard. Even with its low pricing, the iPhone XR still comes packed with lots of features and abilities.

While the other devices were sold at retail prices of $1,099 and $999, the iPhone XR was launched at $749. This price may be high for android users, but old users of the iPhone saw this as a good deal.

Below are the prices of the iPhone XR with different storage capacities:

- $749 for 64 GB
- $799 for 128 GB
- $899 for 256 GB

On September 20, 2019, Apple launched the iOS 13 along with the iPhone 11 series. This new upgrade brought about several new features to the phones that are compatible with the iOS 13 and this includes the iPhone XR. This book has included all the new settings and features available in the iOS 13 to optimize your iPhone XR performance.

iPhone XR Processor

This device has same A12 bionic chip as the XS and XS Max. You may be wondering why the noise on this chip. When compared to all the processors Apple has used from inception, the A12 bionic processor is still their best decision so far. This processor has several processing cores at extra high energy levels which allows you to carry out intensive tasks on your device. Apple says that the A12 bionic chip has been designed to power a minimum of 5 trillion operations per second.

iPhone XR Battery

The iPhone XR has a battery size of 2,942 mAh while the iPhone XS runs on 2,658 mAh. So, for less amount paid for the iPhone XR, you get to enjoy several features and functions.

According to reports by Apple, below are the iPhone XR battery specifications:

- Up to 65 hours audio playback on wireless

- Up to 25 hours talk time on wireless
- Up to 16 hours video playback on wireless
- Up to 15 hours internet use

Apple also gave the bonus of wireless charging on the iPhone XR. This device gives you an extra one and half hour battery life against the iPhone 8.

iPhone XR Display

Several users had anticipated that the XR would have the popular OLED display, however, Apple chose to go with the LCD screen on this one.

Although the LCD screen may be thought as something of the past, Apple prefers to see it as something from the future. With the LCD, the iPhone XR is the first device to have the entire front face of the camera covered. This is why the iPhone XR is tagged the "Liquid Retina Display." This is simply to say that Apple have removed the chin and forehead design that most users know for a long time.

The Liquid Retina Display is also Apple's first LCD device that has the Tap to Wake ability. It also supports the True Tone for adjustments based on lighting in your environment.

Exclusive Wallpapers

On each iPhone XR, you have the pre-loaded custom wallpapers that are designed to match with the exterior of the device. These wallpapers are beautiful and are excusive to the iPhone XR.

Haptic Touch

Apple has removed the 3D Touch on this device and replaced it with “Haptic Touch” which gives similar experience as the 3D touch. To use this feature, just press any applicable element for some seconds until you feel some vibration. With this, you have performed a similar 3D touch action.

There are some limitations on this like not being able to use Haptic Touch on Home Screen icons when you wish to view actions and widgets, but you can access widgets from the control center.

Apple has however promised to improve on this feature.

Dust and Water Resistance

The iPhone XR does not get affected by water and dust as the phone is built to withstand being in water up to one meter deep for as long as 30 minutes. However,

Apple has warned that this is not permanent as normal wear and tear ca reduce the phone's ability to resist water and dust.

Features of the iOS 13

Dark Mode

Every user that has upgraded to the iOS 13 would have the dark mode feature on their phone. Dark mode places a dark background and highlights generally on the phone iOS and all the Apple default apps like messages, mails and so on. Apart from that, this mode has been offered to developers, so with time, the feature would be available in third party apps. You can permanently enable this mode or you can modify it to come on at a particular time or period. In later part of this book, you would see how to achieve this.

HomeKit

HomeKit is also having its own additions. Apart from the small changes to the UI, we have the addition of the HomeKit secure video. Some security cameras that are

HomeKit enabled will now be stored to the HomeKit in a very deep level. This would allow you to store your video recording in iCloud as well as be able to control when exactly you want the camera to record footages.

Routers are not left out as Apple aims to give some security controls and privacy to routers. Apple announced that routers would now be supported in the HomeKit and you can now use the Home app to control the services and devices that can communicate with your router. Not all routers will support this feature though.

Airplay 2 speakers too will now get some more control. You can now use the Airplay 2 speakers to play radio stations, playlists and specific songs through scenes and automatons.

Messages

The message app is not left out of the upgrade. You can now automatically share your profile picture and name with people when you begin a conversation with them.

You can set this option to apply to everyone or just with selected people. This guide would show you how to enable the feature. It is now also very easy for you to search in messages. Simply click on the search bar in the message app and you would see recent messages, photos, people and other options. As you type, the results get refined.

Accessibility

I can say that the biggest accessibility change that we see in the iOS 13 is the voice control addition which would make it easy for you to control your phone with your voice. The system is designed to use the Siri speech recognition algorithm to confirm that it is correct while allowing users to include custom words. Personal data is also kept secure and safe as it leverages on device processing.

Improvement to FaceTime

With the iOS 13, we see some FaceTime video calls improvement, particularly positioning the eye while on calls. FaceTime in iOS 13 will use ARKit to scan your face and softly modify the position of your eyes to make it seem as if you are directly looking into the eyes of your caller rather than at the screen. Although it's small, but it is very impressive.

Privacy

Apple is always known to be particularly with privacy of its users and this was not left out in the iOS 13. Thanks to the new granular controls, you now have control over the location data that each app has access to. When you launch an app, you can choose to give the app access to your location only once or for each time you use the app. Apps would no longer be able to access your location though Bluetooth and Wi-fi. You also now have greater control over the location data on your photos. When you share your pictures, you can choose whether you want the receiver to be able to view the location of the picture or not.

Siri

Siri can now read out incoming messages using the Airpods and you can give an instant response. This function would work without the wake word and is also available with third party messaging apps that has the Sirikit inbuilt. You also have the ***Share Audio Experience*** that allows you to quickly share music with your friends through the Airpods.

For the HomePod, once you get back home, Siri would stop playing music when you tap your phone. Siri now has access to 100,000 radio stations all over the world through the Live radio feature and it also has a sleep timer feature.

Lastly, you can use the Suggested Automations located in the Siri shortcuts to quickly create your personal routines.

Find My

The **Find My Friend** and **Find My iPhone** features were combined into one app in the iOS 13, called the **Find My** app. This app allows you to discover your devices marked as lost even if they are offline by using Bluetooth.

Warnings for Active Subscription

When you delete an app with active subscription, done via in-app purchase on your Apple store, you would get a warning before its deleted.

Animoji and Memoji

The Animojis and Memojis now have more customization options including glasses, makeup, hats, jewelry as well as changing the look of your teeth. Your character can also make use of Airpods. You can also access default Memoji stickers through your device keyboard that you can share across apps like WeChat, Messages and Mails.

Photos and Cameras

We now have a new lighting effect with the portrait camera mode called the **High Key Mono**. You can also change the intensity of the light using the editing features which would make the skin to appear smooth and brighten the eyes. This is similar to the moving lights found in a studio.

The image editing feature available in the photo's app has a swipe-based control as well as a new design. Now, most settings you use for stills can now be used to edit your videos. You can now rotate videos as well as add Filters. The photo library is now designed to delete duplicate photos while it concentrates on what Apple would regard as your best shots. They are then organized by months, day or year and you can see them in a new album view.

Live videos and photos would now auto play in the tab for new photos and you can also see the event, location or holiday etc. in that same tab. The photo tab also has a new birthday mode that gives you access to view the photos of people on their birthday. All your screen recording would now be grouped into a single place so that you do not have to go far to find them.

Sign in with Apple

This is an easy and fast way to sign in to services and apps without having to input your social media login details. This feature authenticates your sign-ins using your Apple ID while ensuring that your personal details are not revealed. This new service uses the TouchID and FaceID as well as a 2-factor authentication system. If the app you want to access require you to verify via email, Apple will generate a random email address that they will forward to your main email just to keep your real details private.

Apple Maps

The Apple maps experienced several alterations and additions including addition of beaches, roads, buildings along with heightened details. The map also has a new feature for favorites on your main screen as well as a collection menu to help you organize your favorites and plan trips. There is also the introduction of the **Look Around** which is similar to Google's **Streetview** to give you a view of a location before you visit. In this guide, you would learn how to use each feature for the map.

Other additions include being able to share your ETA with your family and friends, get updated flight information, real time updates on public Transportation as well as an improved Siri navigation.

CarPlay

The CarPlay too had its own revised interface. You now have a main screen called the Dashboard where you would see controls for podcast or music playback and also see basic maps information and have control on some HomeKit devices like garage door opener. Other new features include a new calendar app to show a quick view of the day when embarking on a journey. Apple music is also modified to make it easy to find music.

Reminders

Reminders now have the toolbar to add in times, dates and attachments. Also, there has been a great improvement with the message app integration to allow both apps communicate together. This means that you can set a reminder about a contact and the reminders would pop up when messaging the contact.

Swipe Typing

With the introduction of the new dark mode came the swipe typing feature. You would find in details how this works later in the book.

Health App

All the functions that is on the new Watch OS health app has now been replicated on iPhone like the menstrual cycle tracking and the activity trends. The main app also has a new summary view to display notifications and a highlight section where you would see your health and fitness data overtime. The app uses machine learning to display things that are of utmost importance to you. It is either encrypted securely on iCloud or stored on your phone and you can decide to share any of the details with other people

CHAPTER 1

Getting Started: How to set up your iPhone XR

Setting up your device is the first and most important step to getting started with your iPhone XR. Follow these steps for a seamless experience.

1. Firstly, you need to power on your device. To do this, press and hold the side button. Now, you will see **"Hello"** in various languages. The screen would present options to set up your device. Follow the options presented on the screen of your device.

   Note: From the Hello screen, you can activate the **Voice Over or Zoom Option** which is helpful for the blind or those with low vision.
2. A prompt would come up next to select your language and country/ region. It is important you select the right information as this would affect how information like date and time etc. is presented on the device.
3. Next is to manually set up your iPhone XR by tapping "**Set up Manually**". You can choose the "**Quick Start"** option if you own another iOS 11 or

later device by following the onscreen instruction. If you don't have this, then set up your iPhone manually.

4. Now, you have to connect your device to a cellular or Wi-Fi network or iTunes to activate your phone and continue with the setup. You should have inserted the SIM card before turning on the phone if going with the cellular network option. To connect to a Wi-Fi network, just tap the name of your Wi-fi and it connects automatically if there is no password on the Wi-fi. If there is a security lock on the Wi-fi, the screen would prompt you for the password before it connects.
5. At this stage, you can turn on the Location services option to give access to apps like **Maps** and **Find my Friends**. This option can be turned off whenever you want. You would see how to turn on the location services and how to turn it off completely on your iPhone in a later part of this book.
6. Next is to set up your Face ID. The face ID feature gives you access to authorize purchases and

unlock your devices. To setup the Face ID now, tap Continue and follow the instructions on the screen. You can push this to a later time by selecting "**Set Up Later in Settings.**"

7. Whether you setup Face ID now or later, you would be required to create a four-digit passcode to safeguard your data. This passcode is needed to access Face ID and Apple Pay. Tap **"Passcode Option"** if you would rather set up a four-digit passcode, custom passcode or even no passcode.
8. If you have an existing iTunes or iCloud backup, or even an Android device, you can restore the backed-up data to your new phone or move data from the old phone to the new iPhone. To restore using iCloud, choose "**Restore from iCloud Backup"** or "**Restore from iTunes Backup"** to restore from iTunes to your new iPhone XR. In the absence of any backup or if this is your first device then select "**Set Up as New iPhone".**
9. To continue, you would need to enter your Apple ID. If you have an existing Apple account, just enter the ID and password to sign in. In case you don't have an existing Apple ID or may have

forgotten the login details, then select **Don't have an Apple ID or forget it.** If you belong to the class that have multiple Apple ID, then select **Use different Apple IDs for iCloud & iTunes** on the screen of the phone.

10. To proceed, you need to accept the iOS terms and conditions.
11. Next is to set up Siri and other services needed on your device. Siri needs to learn your voice so you would need to speak few words to Siri at this point. You can also set up the iCloud keychain and Apple Pay at this point.
12. Set up screen time. This would let you know the amount of time you spend on your device. You can also set time limits for your daily app usage.
13. Now turn on automatic update and other important features.
14. Click on "**Get Started**" to complete the process. And now, you can explore and enjoy your device.

How to Insert SIM in iPhone XR

Before you can use your device, you must have inserted the SIM. To do this, please follow the steps below:

- The iPhone pack comes with an opener used to open the SIM and memory card holder.
- Put the opener into the tiny hole in the SIM holder.
- Pull out the SIM holder once the opener is able to clip it out.
- Set your SIM to ensure that the angled corner of the SIM is placed in the angled corner of the SIM holder. Note that iPhone XR only support Nano SIMs.
- Push in the SIM holder once the SIM has been correctly placed.
- Now your SIM is ready to be used.

How to Charge the Battery for the iPhone XR

It is important that you charge your phone often to ensure its ready for use at all times.

- Connect the phone charger to a power socket and then connect the USB side to the lightning port at the bottom of the phone.

- To know that your battery is charging, you would see the battery charging icon displayed at the top of the screen.
- At the top right side of the screen, you would see your battery level. The more the colored section, the more power the device have and vice versa.

How to turn on iPhone XR

The following steps would show how to turn on the iPhone XR:

- Press the **Side** button until the iPhone comes on.
- Once you see the Apple logo, release the button and allow your iPhone to reboot for about 30 seconds.
- Once the iPhone is up, you would be required to input your password if you have one.

How to turn off iPhone XR

Follow the steps below to turn off your iPhone XR:

- Hold both the volume down and the side button at same time.

- Release the buttons once you see the power off slider.
- Move the slider to the right for the phone to go off.
- You may also use the side and volume up button, only thing is, you may take a screenshot in error rather than shutting down the phone.

How to Download iOS 13 on iPhone

To be able to enjoy the features packed in the iOS 13, you first have to download it on your phone. First step is to ensure that your iPhone device has been backed up to make it easy for you to restore your device in case you lose your contents during the upgrade.

How to Backup Using iCloud

This is probably the simplest way to back up your device with the steps below

- Connect your device to a Wi-Fi network.
- Go to the settings app.

- Click on your name at the top of the screen.
- Then click on iCloud.
- If using iOS 10.2 or earlier version, navigate to **iCloud** on the settings page
- Navigate down and click on **iCloud backup**.
- Then select **Back Up Now.**
- If using iOS 10.2 or earlier version, simply click on **backup.**

To check whether the backup was completed, follow the steps below

- Go to settings.
- Click on **iCloud.**
- Then select **iCloud Storage.**

- And choose **Manage Storage** on the next screen.
- Then click on your device from the list.

How to Back Up on MacOS Catalina

Although the MacOS 10.15 Catalina no longer has the iTunes icon as they replaced it with apps for Podcasts, Books and Music, you can still use the new MacOS to back up your iPhone device.

- Connect your iPhone device to the Mac and ensure its updated.
- Follow the instructions on the screen and enter your passcode if requested or activate the ***Trust This Computer*** option.
- Launch the **Finder App.**
- Choose your iPhone device from the side bar.
- Click on **General.**
- Then click on ***Back Up Now*** *to begin a manual backup.*

How to Back Up with iTunes on PC or Mac

if you have a Windows PC or an older Mac, you can use the iTunes to back up your iPhone with the steps below:

- Confirm that the iTunes is updated to the current version then connect your iPhone.
- Follow the instructions on the screen and enter your passcode if requested or activate the ***Trust This Computer*** option.
- In the iTunes app, click on your iPhone.
- Then click on ***Back Up Now*** to back up your device.

How to Download and Install iOS 13 on your iPhone

The best way to download iOS 13 on your iPhone is via the air with the steps below

- Go to the settings app

- Click on **General**
- Then select **Software Update**
- Your device would begin to search for a new update after which you would receive notification of the iOS 13.
- Click on ***Download and Install.***
- It may take a while to download and you would not be able to make use of your device during the update.

How to Download and Install iOS 13 on PC or Mac Through iTunes

If you would rather download the iOS 13 to your PC or Mac through iTunes, use the highlighted steps below:

- You have to be on the most current version of iTunes.
- Connect the iPhone to your computer.
- Launch iTunes then click on your iPhone.
- On the next screen, select **Summary.**
- Then click on ***Check for Update.***

- *Finally click on* ***Download and Update.***

How to Use Cycle Tracking in Health

The health app in the Apple device is a great tool and the iOS 13 has brought more additions, the biggest of this addition is the cycle tracking. With this tool, you can track your menstrual cycle and also have tools to alert you when you are at your most fertile days and when you are due. To get started, follow the steps below

- Launch the health app.
- Then choose **Search**, and click on **Cycle tracking** from the displayed list.
- Click on **Get started**, then click on **Next.**

- The app would ask you series of questions like duration of your period and the date the last one started.

- You would also get options on ways you would like to track your period; you can also choose to be given notifications and predictions on when you are likely to see the next cycle. You would also select whether you would like to record spotting and symptoms in your cycle log and whether you want to be able to view your fertility Windows.

- As soon as you have provided all the answers, you would return to the homepage for **Cycle Tracking.**

- On this screen, click om **Add Period** to choose days that you have experienced your periods.

- You can also click on Spotting, **Symptoms** and Flow Levels option to input more specific details.

Going Home on your iPhone XR

- Regardless of where you are on your iPhone XR, to return to the home screen, simply swipe the screen from the bottom up.

How to Choose Ringtone on the iPhone XR

- From the Home screen, go to **Settings.**
- Click on **Sounds & Haptics.**
- Then click on **Ringtone.**
- You may click on each of the ringtones to play so you can choose the one you prefer.
- Select the one you like then click the "**< Back**" key at the top left of the screen.
- Slide the page from bottom up to return back to Home screen.

How to Choose Message Tone on the iPhone XR

- From the Home screen, go to **Settings.**
- Click on **Sounds & Haptics.**
- Then click on "**Text Tone**".

- You may click on each of the message tones to play so you can choose the one you prefer.
- Select the one you like then click the "**< Back**" key at the top left of the screen.
- Slide the page from bottom up to return back to Home screen.

How to Use the Find My App

In the iOS 13, Apple combined the Find My iPhone and Find My Friends feature into an app they called **Find My.** With this feature you can share your location to your loved ones and friends as well as find your devices using the same app. It is simple to use with the steps below

- Go to your home page and launch the **Find My** app.

- Under the **People** tab, you would sew your current location.

- Click on the **Start Sharing Location** tab to share with a contact.

- Type in your desired contact to share with.

To find your missing device,

- Click on the **Device** tab to modify your map to present all the registered Apple devices on your account.

- Click on the missing device then select from any of the options on the screen: ***Mark As Lost*, Get Directions** to the device, **remotely *Erase This Device,*** or ***Play Sound.***

- *If any of the device is currently offline, you can set the map to alert you once the device is connected to the internet. Simply click the **Notify Me** option.*

How to Edit Photos and Rotate Videos

Other iOS had lacked its own editing tools until the iOS 13. Now, you can go to your photo app to modify various key parts of tour videos and photos. To do this,

- Go to your photos app.

- Choose your desired photo and then click on the **Edit** button.

- On the next screen, you can try to swipe between options and adjust sliders to see how your picture would look when modified with several options

- You can also do this for the videos.

How to Use Sign-IN-With-Apple

It can be quite tiring having to log in each time you launch a different app and at the same time you may not be comfortable signing in with your Instagram account to all apps. The Sign in with Apple allows you to quickly sign into apps with your Apple account while protecting the need to not share personal information.

- For apps that support this feature, you would see the option displayed on the opening screen of the apps.
- Click on it and you would be prompted to login to your Apple account
- Then you would select the information you desire to share with the app developer.
- With this feature, you can decide to share your email address or not to share it.

- If you do not want to share your personal email address, Apple would generate a random email address that would automatically forward to your Apple iCloud email address while keeping your anonymity safe.

How to Set/ Change Language on iPhone XR

- From the Home screen, click on the **Settings** option.
- Select **General** on the next screen.
- Then click on **Language and Region.**
- Click on **iPhone Language** to give you options of available languages.
- Choose your language from the drop-down list and tap **Done.**
- You would see a pop-up on the device screen to confirm your choice. Click on **Change to (Selected Language)** and you are done!

How to Set/ Change Date/ Time on iPhone XR

- From the Home screen, click on the **Settings** option.
- Select **General** on the next screen.

- Then click on **Date & Time.**
- On the next screen, beside the **"Set Automatically"** option, move the switch right to turn it on.
- Slide the page from bottom up to return back to Home screen.

How to Use the Control Centre

- From the right side of the notch, swipe down to view the control center.
- Click on the needed function to either access it or turn it on or off.
- Move your finger up on the needed function to choose the required settings.
- Once done, return to home screen.

How to Choose Settings for the Control Centre

- Go to Settings> Control Centre.
- On the next screen, beside the **Access within Apps** option, move the switch to turn it on or off.
- Click on **Customize Controls.**
- For each function you want to remove, click on the minus (-) sign.

- To add icon under **More Controls,** click on the plus (+) sign at the left of each of the icons you want to add.
- Click on the move icon beside each function and drag the function to the desired position in the control centre.
- And you are done.

How to Connect to Paired Bluetooth Devices from Control Center

It is now easier to access paired devices on iOS 13. As is usual, clicking on the Bluetooth button would either enable or disable Bluetooth. Now the iOS 13 has added the 3D Touch feature to display devices and also connect. With this new addition, you do not need to exit a current app to look for the settings app and all the long processes available in the other versions. Should you need to pair a device to Bluetooth, you can now do so from the control center without exiting the current app you are on. The steps are highlighted below:

- Launch the control center. If your iPhone has a home button, just swipe from the bottom of your screen up to access the control center. If your iPhone has no home button, swipe from the right top side of the iPhone down to access control center.

- You either 3D touch or you click and hold the wireless connections block at the top right side of the screen to expand it.

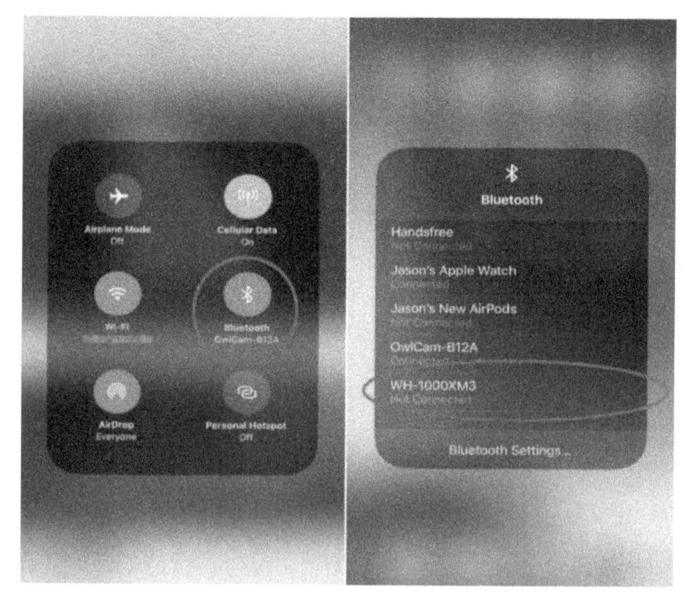

- 3D touch or Tap and hold the Bluetooth button at the right of the screen.
- You would see a list of all the Bluetooth devices that have been paired whether connected or not.
- Select the one you wish to connect to and you are fine.

How to Quickly Connect to Wi-Fi on iOS 13 Through the Control Center

Most of us are used to joining new Wi-fi networks very frequently whether at a friend's place, a restaurant or while on a flight. The iOS 13 has now made it easier to connect. Rather than launching the settings app to be able to view the Wi-fi menu, you can now connect directly from the control center. Again, you would not need to exit an app to do this. See the steps below.

- Launch the control center. If your iPhone has a home button, just swipe from the bottom of your screen up to access the control center. If your iPhone has no home button, swipe from the right top side of the iPhone down to access control center.

- You either 3D touch or you click and hold the wireless connections block at the top right side of the screen to expand it.

- 3D touch or Tap and hold the Wi-Fi button at the left side of your screen.

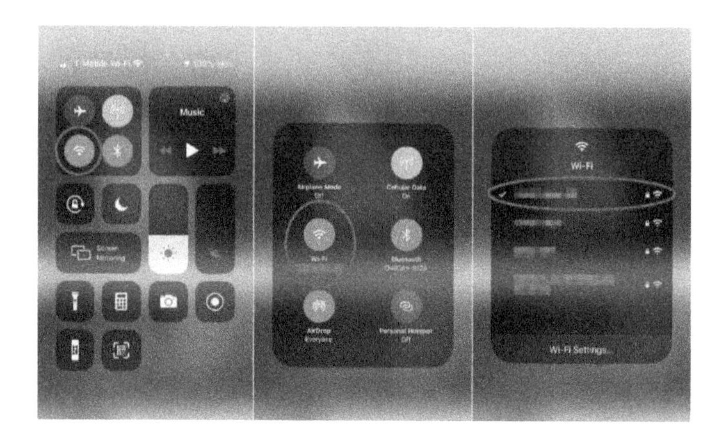

- You would see a list of all nearby Wi-fi networks that have been paired whether connected or not.

- Select the one you wish to connect to and you are fine.

How to set up Apple ID on iPhone XR

- Go to the **Settings** option.
- At the top of your screen, click on **Sign in to your iPhone.**
- Choose **Don't have an Apple ID or forgot it?**
- A pop-up would appear on the screen and you click on **Create Apple ID**.
- Input your date of birth and click on Next.

- Then input your first name and last name then click on **Next.**
- The next screen would present you with the email address option. Click on **"Use your current email address"** if you want to use an existing email or click on **"Get a free iCloud email address"** if you want to create a new email.
- If using an existing email address, click on it and input your email address and password.
- If creating a new one, click on the option and put your preferred email and password.
- Verify the new password.
- Next option is to select 3 Security Questions from the list and provide answers.
- You have to agree to the device's Terms and conditions to proceed.
- Select either **Merge** or **Don't Merge** to sync the data saved on iCloud from reminders, Safari, calendars and contacts.
- Click on **OK** to confirm the **Find My iPhone is turned on.**

How to Set Up Apple Pay

- First is to add your card, either debit, credit or prepaid cards to your iPhone.
- To use the Apple Pay, your device should be updated to the latest iOS version.
- You should be signed into the iCloud using your Apple ID.
- To use the Apple Pay account on multiple devices, add your card to each of the devices.

To add your card to Apple Pay, do the following:

- Go to Wallet and click on 
- Follow the instructions on the screen to add a new card. On iPhone XR, you can add as much as 12 cards. You may be asked to add cards that is linked to your iTunes, cards you have active on other devices or cards that you removed recently. Chose the cards that fall into the requested categories and then input the security code for each card. You may also need to download an app from your card issuer or bank to add your cards to the wallet.
- When you select **Next,** the information you inputted would go through your bank or card

issuer to verify and confirm if the card can be used on Apple Pay. Your bank would contact you if they need further information to verify the card.

- After the card is verified, click Next to begin using Apple Pay.

How to check out with Apple Pay

Here are useful steps to check out on Apple Pay for your daily transactions:

- To make a payment at a checkout terminal, double-press the side button to open the Apple Pay screen.
- Look at the iPhone screen to verify with Face ID (or enter your passcode).
- And then place the iPhone XR near the payment terminal.
- If you're using Apple Pay Cash, double-press the side button to approve the payment.

How to use Siri on iPhone XR

Apples' virtual assistant is a delight to work with, everyone loves Siri, and most of the time you spend with her involves getting an answer, but she can do more than answer questions.

1. **How to Set up Siri on iPhone XR**

To use Siri on your iPhone XR, you have to set it up like you set up the Face ID. Find below the steps to do this on your iPhone XR.

- Click on **Siri & Search** from the **Settings** app.
- Beside the option "**Press Side Button for Siri",** move the switch to the right to enable the function.
- A pop-up notification would appear on the screen, select "**Enable Siri".**
- Switch on the "**Listen to Hey Siri"** option and follow the instructions you see on the screen of your iPhone. (To use Siri when your phone is locked, activate the **Allow Siri When Locked** option).
- Click on **language** and select the desired language.

- Click on the **< Siri & Search** button at the top left of the screen to go back.
- Scroll and select **Siri Voice.**
- On the next screen, select accent and gender.
- Click on the **< Siri & Search** button at the top left of the screen to go back.
- Select "**Voice Feedback**".
- Choose your preferred setting.
- Click on **< Siri & Search** at the top left of the screen.
- Select **My Information.**
- Click on the contact of choice. If you set yourself as the owner of the phone, the device would use your data for various voice control functions like navigating home. You can create yourself on the contact by following the steps given in creating contact.
- Select the desired application.
- Next to **"Search and Siri Suggestions"** slight left or right to turn on or off.

Now Siri is set up and ready to be used.

2. **How to Activate Siri on the iPhone XR**

There are 2 ways to activate Siri on your iPhone XR.

- Voice option. If you enabled “Hey Siri”, then you can begin by saying “Hey Siri” and then ask Siri any question.
- Using the side button. To wake Siri, press the side button and ask your questions. Once you release the side button, Siri stops listening.

3. **How to Exit Siri**

To exit Siri, follow the simple step below.

- Press the side button or swipe u from the bottom of the display to exit Siri.

How to Enable Dark Mode

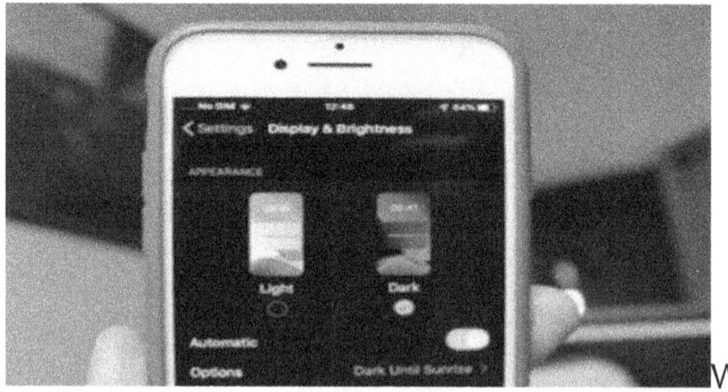

We wake up in the morning, eager to see our missed notifications, you pick your phone and get almost blinded by the bright white theme of your phone. Thankfully, the iOS 1e comes with a new dark mode that can save you from this brightness by changing the white areas in your phone to a much darker tone. This change would apply to all the system apps including Safari and iMessages while Apple has encouraged all third-party developers to add themes that are compatible with the dark mode in their apps. Follow the steps below to find Dark Mode

- Go to the settings app
- Then click on **Display and Brightness**

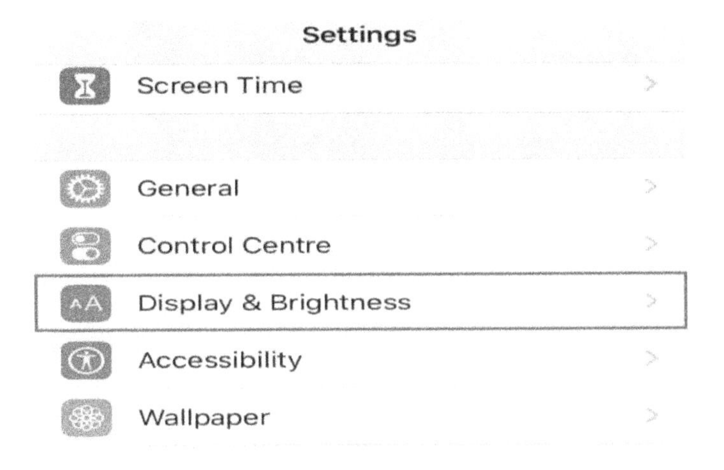

- Then tap on the tick box beneath the Light or Dark themes to activate your preferred one.

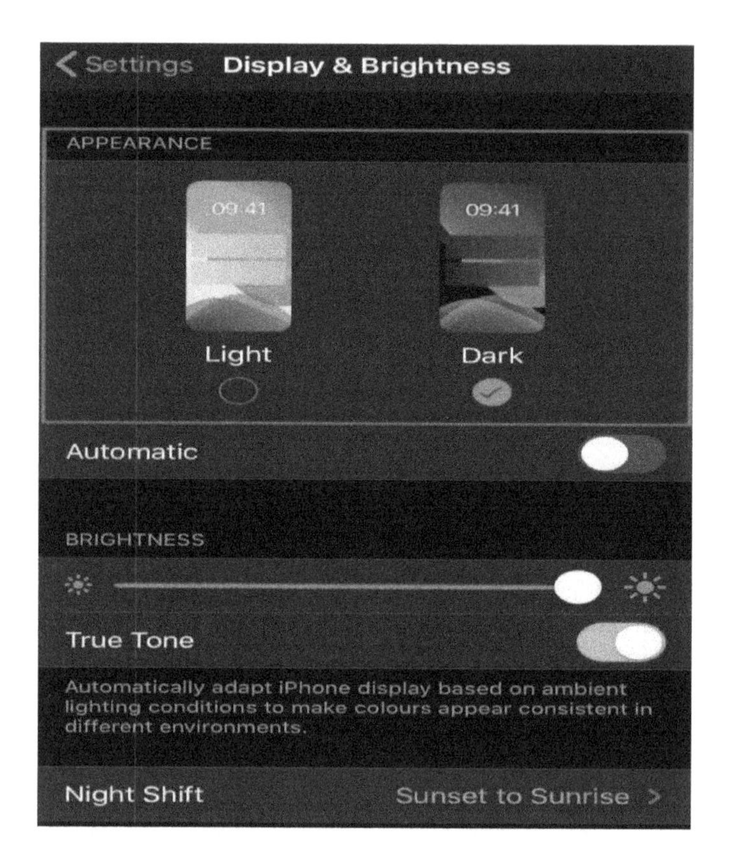

- If you want to set your phone to have a bright theme in the day and a dark theme at night, just hit the toggle for the **Automatic** option and tap the **Options** button under **Automatic** to set when the darker theme should set in.
- Select either **Sunset to Sunrise** option or set to **Custom Schedule.**

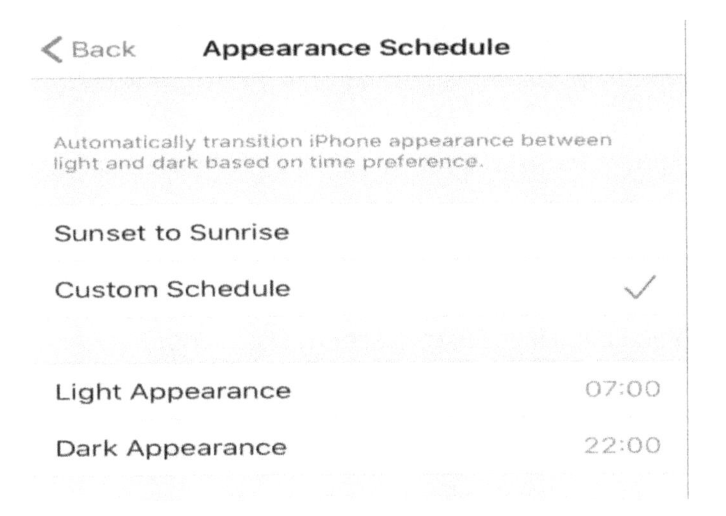

How to Automatically Activate the Dark Mode

If you wish to use a brighter theme during the day and a darker theme at night, you can do this without having to always go to settings every time to configure. You can configure your settings to interchange the two options at the set time. You can do this with the simple steps below:

- Go to the settings app
- Then click on **Display and Brightness**
- Beside the **Automatic** menu, move the switch to the right to enable it.
- The menu would by default change to **Sunset to Sunrise.**

- To edit this, click on **Options** under the **Automatic** menu.
- **Sunset to Sunrise** means that Dark Mode would be activated once the sun goes down using your GPS location.
- You can select ***Custom Schedule*** and input your own desired time for the Dark Mode to kick in.

How to Set Your Wallpaper to React to Dark Mode

Do you know that some wallpapers in the iOS 13 can react to the Dark Mode? Although small, it can be quite fun. To set a wallpaper that has a dynamic color changing feature, follow the simple steps below:

- Go to the settings app.
- Click on **Wallpapers.**
- Then select ***Choose a New Wallpaper****.*
- *Then chose* ***Stills.***
- *Wallpapers that can react when the Dark Mode is enabled are marked with a bisected small circle at the right of your screen, towards the bottom and you would see a line down the image middle to display what changes you would get if activated.*

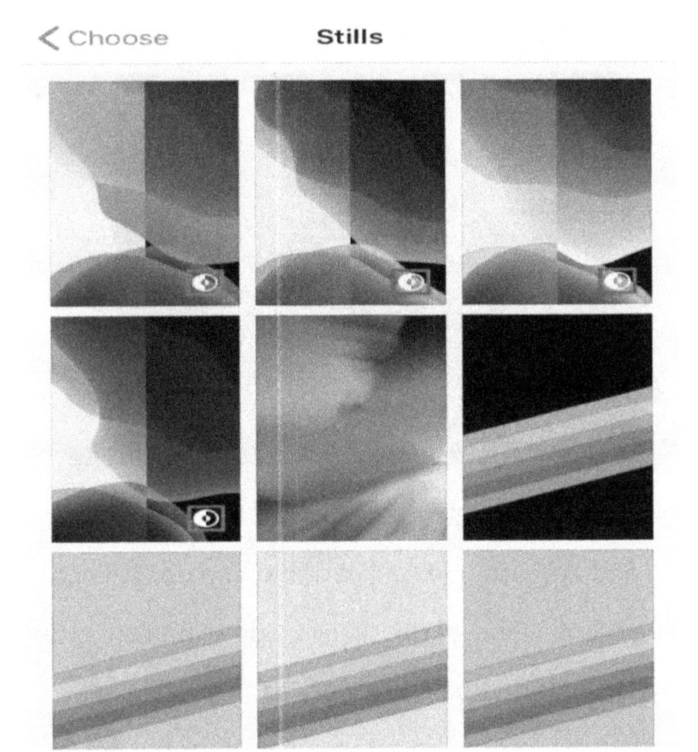

- If you would rather have your own customized wallpaper, return to the **Wallpaper menu**
- Navigate to ***Dark Appearance Dims Wallpaper*** and toggle the switch to the right to enable. Although the wallpapers would not react like the reacting ones, however, it would dim a little when the Dark mode is enabled so that you don't get dazzled by the lighter areas.

How to Tap and Drag the New Volume Indicator

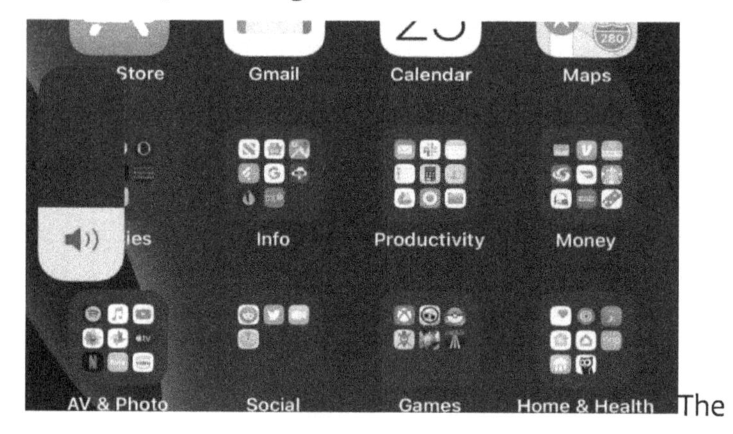

The new volume indicator introduced in the iOS 13 is less obtrusive and you can now pull it down and up. Several users have complained of the giant volume indicator for years now and Apple finally used a small vertical bar placed at the top left of the screen to replace it. Use your finger to drag the bar down and up. When you do this, an indicator would show at the end of the volume bar showing the output of your device sound, for example, Bluetooth device Airpods or speaker.

How to Download Large Apps over Cellular Network

You no longer have to wait for Wi-fi connection to be able to download large apps. Before now, your iOS would always warn you to connect to wi-fi to download large files but with the iOS 13, when downloading apps over 200MB, you would receive a pop up on your screen asking if you want to download with cellular or if you would rather wait for Wi-fi.

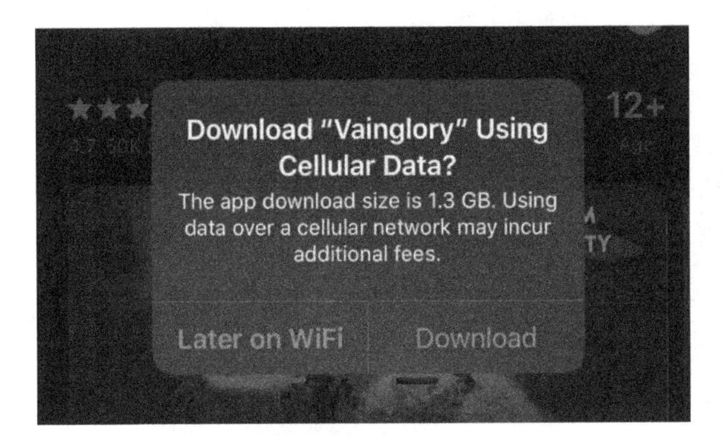

- Go to the settings app.
- Click on **iTunes & App Store.**
- Here, you can set the iOS to always allow you to download your apps over cellular or always ask you if you want to continue or ask you only when the app is over 200MB.

How to Set Optimized Battery Charging

With the iOS 13, you can optimize your battery charging to enable it last longer with the steps below:

- Launch the settings app.

- Click on **Battery.**
- Navigate to **Battery Health.**

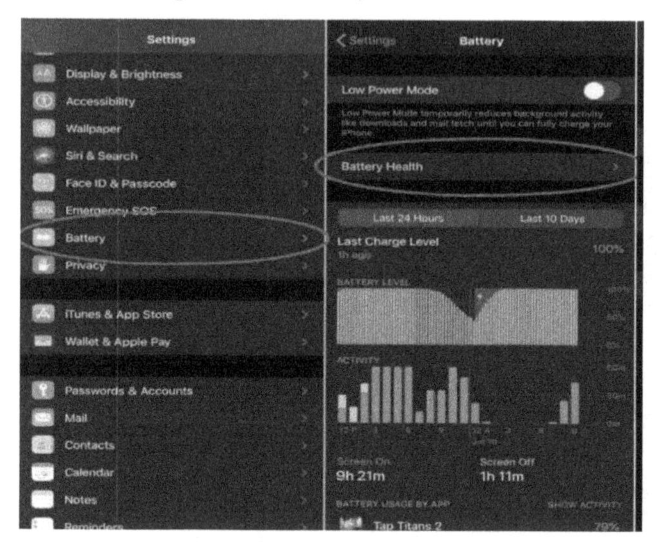

- Under **Battery Health,** you would find the maximum battery capacity left for your battery, an indication of the level of degradation as well as the option to enable Optimized Battery Charging.
- Move the switch beside **Optimized Battery Charging** to enable the feature.

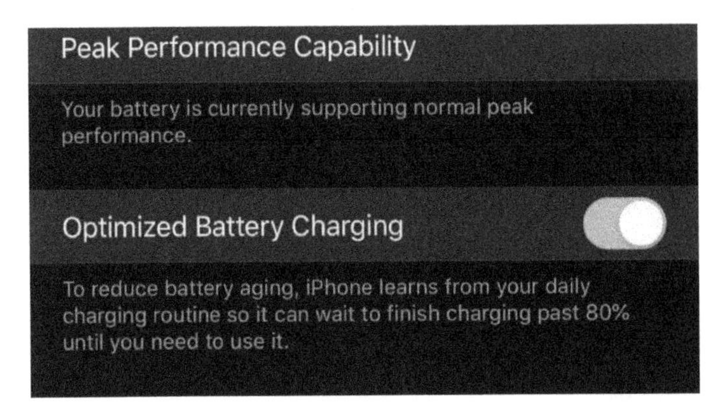

Note: while this feature is on, you may notice that your iPhone could stop charging at 80%. You should not worry as Apple has ensured that the device does not drain battery quick.

Other Tips to Improve the Longevity of your iPhone Battery

Several things work together to reduce the life of your battery like plugging the device even after it is fully charged. Below I would list tips that would help to prolong the life of the battery.

- Its important not to let your battery to totally get drained. Best is to always keep it at 0 percent and above.

- Avoid exposing the device to excessive heat. It is a very bad idea to charge your iPhone in a very hot environment.

- Quick change from very hot to very cold condition is also very bad for the health of your battery.

- If you intend to not use your phone for a week and above, it is advisable you run down the battery to below 80% but not less than 30 percent. Then shut down the phone completely.

- Do not always fully charge your device when not using it for a long time.

How to Pair your iPhone with a DualShock 4

You can play games on your phone using the DualShock 4 for a better experience. Follow the steps below to pair both devices.

- Go to the settings app.

- Click on **Bluetooth** to enable it.

- With the Bluetooth enabled, ensure that the DualShock 4 controller is well charged.

- Press both the share button and the PlayStation button at same time and hold down for some seconds.

- Then you would see the light at the back of the controller begin to flash intermittently.

- Under the Bluetooth menu on your iPhone, you would see DualShock 4 Wireless controller come up as one of the devices.

- Click on it.

- The blinking light at the back of the controller should change to reddish pink color as an indicator that the devices are paired.

How to Disconnect a DualShock 4 from your iPhone

After you are done with the game, it is advisable to turn off the Bluetooth connection. To disconnect through the controller, just hold down the PlayStation button for approximately 10 seconds.

To disconnect from your iPhone, the best method would be to go through the Control Center with the steps below

- If using a Face ID compatible iPhone, launch the control center by swiping diagonally from the top right to the lower left of your screen.

- If using a Touch ID compatible iPhone, use your finger to swipe from the bottom up.

- Hold down the Bluetooth icon on your screen.

- A menu would pop up on your screen, hold down the icon for *Bluetooth: On*.

- Another pop up would show on the screen displaying "DUALSHOCK 4 Wireless Controller" in the list.

- Click on it to disconnect your controller.

Another method you can use is below

- Go to the settings app and click on **Bluetooth.**

- A pop up would appear on your screen, hold down the icon for *Bluetooth: On*.

- Another pop up would show on the screen displaying "DUALSHOCK 4 Wireless Controller" in the list.

- Click on it to disconnect your controller.

And again, another method can be found below

- Go to the settings app and click on **Bluetooth.**

- On the next screen, under the **My Devices** List, you would find "DUALSHOCK 4 Wireless Controller"

- At the right of this option, you would find an icon with "I" in a blue circle. Click on this icon.

- A menu would pop up, then select **Disconnect.**

When next you want to use the controller, simply press the PlayStation button to immediately connect.

How to Unpair the DualShock 4 from your iPhone

Usually, the PlayStation may connect by accident when stuffed in your bag especially when on a trip. In such cases, it is advised to unpair first and then re-pair when you want to use it.

To unpair this device, follow the steps highlighted above to disconnect but rather then clicking on **Disconnect,** you should click on **"Forget This Device"**

How to Pair your iPhone with an Xbox One S controller

- Go to the settings app.
- Click on **Bluetooth** to enable it.
- With the Bluetooth enabled, ensure that the **Xbox One controller** is well charged.
- Press the Xbox logo button to turn on the Xbox.

- You would see a wireless enrollment button located at the back of the controller. Press the button and hold it for some seconds.

- If the controller has been unpaired already from a different device, please skip this step as you can just press and hold the Xbox button to put it in pairing mode.

- Then you would see the light of the Xbox button begin to flash quickly.

- Under the Bluetooth menu on your iPhone, you would see “Xbox Wireless Controller” come up as one of the devices.

- Click on it.

- Once it is paired correctly, the blinking light would stop and remain focused.

How to Disconnect Xbox One Controller from your iPhone

After you are done with the game, it is advisable to turn off the Bluetooth connection. To disconnect through the controller, just hold down the Xbox button for approximately 10 seconds.

To disconnect from your iPhone, the best method would be to go through the Control Center with the steps below

- If using a Face ID compatible iPhone, launch the control center by swiping diagonally from the top right to the lower left of your screen.

- If using a Touch ID compatible iPhone, use your finger to swipe from the bottom up.

- Hold down the Bluetooth icon on your screen.

- A menu would pop up on your screen, hold down the icon for *Bluetooth: On.*

- Another pop up would show on the screen displaying “Xbox Wireless Controller” in the list.

- Click on it to disconnect your controller.

And another method can be found below

- Go to the settings app and click on **Bluetooth.**

- On the next screen, under the **My Devices** List, you would find “Xbox Wireless Controller.”

- At the right of this option, you would find an icon with “I” in a blue circle. Click on this icon.

- A menu would pop up, then select **Disconnect**.

To use this device again, simply press the Xbox button to get it working.

How to Unpair the Xbox Controller from your iPhone

Usually, the controller may connect by accident when stuffed in your bag especially when on a trip. In such cases, it is advised to unpair first and then re-pair when you want to use it.

To unpair this device, follow the steps highlighted above to disconnect but rather then clicking on **Disconnect,** you should click on **"Forget This Device"**

How to Download Fonts from the APP store

Fonts that you download from the app store are downloaded as apps. After downloading, you would see them in the formatting menu when using apps that support the downloaded fonts like Notes, emails etc. Follow the guide below to download fonts to your device.

- Go to the app store.
- Search for fonts by typing 'Fonts for iPhone' in the search bar.
- Click on **GET** when you find a fonts app.
- For users that may have enabled Passcode, Face ID, or Touch ID on iTunes & App Store when installing apps, Apple would ask that you register for one of the 3 options before you can then install the fonts.
- After the Fonts app has been installed, click on **OPEN.**

- Give the required permissions to the app.
- Then click on the **'+' icon** at the left top side of your screen.
- Input a name for the collection then click on **OK.**
- On the next screen, click on the **'+' icon** at the left top side of your screen.
- You would see 2 options appear from the bottom: **Font Squirrel** and **Google Fonts.** Select **Google Fonts.**
- The next step is to choose the fonts you wish to add. While you can select multiple fonts, however, it is better to choose less than 10 fonts at a time. Once done, click on **'Add to collection.'**
- Then click on **OK.**
- Use the back arrow on your screen to Go back and then click on **'Install fonts'** at the bottom of your screen.
- This would take you to Safari browser where you would see a dialog box asking you to the configuration.
- Click on **'Allow.'**
- Once the download is completed, you would see a pop-up on your screen saying 'Profile Downloaded.'
- Click on **Close** to exit the app.
- Then go to the settings app on your iPad.
- Click on **General.**
- Then click on **Profile** towards the bottom of the next page.
- Click on the saved name for the downloaded fonts.

- Then click on **Install** at the right top corner of your screen.
- You may be prompted to input passcode, input the code and click on **Install** again.
- Once the installation is complete, click on **Done.**
- Return back to **Settings.**
- Click on **General.**
- Then select **Keyboard.**
- Click on **Keyboard** again on the next screen.
- Then click on **Add New Keyboard'**
- **Navigate to Fonts and click on it and your downloaded fonts are ready for use.**
- **Launch any messaging/ chat apps like the Message app.**
- **Click on your desired conversation or begin a new conversation.**
- **Long press on the Globe icon** at the left bottom corner of your screen, then chose **Fonts** from the list.
- You would see all available fonts above your keyboard.
- Select the font of your choice and begin to chat.

CHAPTER 2: BASIC FUNCTIONS

How to Wake and Sleep Your iPhone XR

Waking and sleeping your iPhone XR will preserve your battery Life and make it long lasting; here are the steps for wake and put your iPhone XR to sleep.

- There are two ways to wake your iPhone XR and to see the lock screen; either tap the screen or just picking up the device and glancing at it can wake the iPhone XR.
- From the lock screen, you can use either the camera or the flashlight by a simple gesture. Tap and hold the individual icon until you hear a click sound.
- Simply press the side button to make the iPhone XR go to sleep. With the Apple Leather Folio case designed for the iPhone XR, simply open the case to wake and close the case to sleep your device.

How to Set up Face ID on iPhone XR

There are several things you can do just by glancing at your device. With Face ID, you can unlock your device, sign into apps, authorize purchases and lots more.

Before setting up the Face ID, ensure nothing is covering your face or the TrueDepth camera. Apple designed the Face ID to work well with contacts and glasses. To get the best result, let your iPhone be about an arm's length from your face (10 – 20 inches or closer). Now, see steps to set up Face ID:

- Visit Settings > Face ID & Passcode. Input your passcode if asked.
- Select **Set Up Face ID.**
- Put your device in portrait orientation and place your face in front of your iPhone and tap "Get Started"
- The screen comes up with a frame, set your face to fit into the frame and move your head slowly until the circle is complete. Tap **Accessibility Options** if you are not able to move your head.
- Once done with the first Face scan, click on continue.
- Move your head slowly again until the circle is completed for the second time.
- Tap **Done** to complete.

- If you do not already have a passcode set, you would be prompted to create one as an alternative option for identity verification.
- Go to **Settings** then **Face ID & Passcode** to activate features to go with the Face ID. This includes iTunes & App Store; iPhone Unlock and Safari AutoFill.

How to Unlock your iPhone XR using Face ID

The wonder of the iPhone XR is the ability to unlock the device with Face ID. Follow the steps to unlock your iPhone.

- Go to **Settings** then **Face ID & Passcode.**
- Go to the option **"Use Face ID For"** and switch on **iPhone Unlock.**
- To unlock your iPhone, wake the device first then you glance into the screen.
- The iPhone XR would automatically scan your face and authorize the login attempt.
- Once successful, the **lock icon** on the phone screen will open.

- To unlock the device, simply swipe from the bottom of the iPhone up to show the home screen.

With the iPhone XR, only a single face is supported on Face ID unlike the Touch ID on other iPhones. You would be unable to create faces of friends and family members on your iPhone. The only way a third party can access your phone is to manually input the password. This option would also work for you if the Face ID isn't working.

How to make Purchases with Face ID on iPhone XR

If the iTunes and App Store are activated for Face ID under **Face ID & Passcode,** you can use the Face ID to carry out purchases on the App store, iTunes Store and iBooks store.

Follow the steps below.

1. Open any of the stores on your phone.
2. Search for the items you want to purchase and click on it.
3. To make payment, click the **Side** button twice and look at your iPhone XR.

4. Once its completed, a message would pop up showing **Done** with a **Checkmark** on your device screen.

How to Transfer Content to your iPhone XR from an Android Phone

You can move contents to your device from an Android mobile phone when you first activate the device or after you did a factory reset. To do this, you would see the **Apps and Data** option on your screen.

- Under **Apps and Data,** click on "**Move Data from Android".**
- You have to install the app "**Move to iOS"** on the android phone before you can move data.
- Click on **Continue** when you have downloaded the app.
- Follow the instructions you see on the screen to move data from the Android to the iPhone XR.

How to Apply Filter to a Video in iOS 13

One of the additions in the iOS 13 is the filter tool. So now you can apply a filter similar to the ones used on Instagram in videos that you have captured already.

- Open the photo apps on your device.
- Choose a video from the photo library.
- Click on **Edit** at the right top side of your screen.

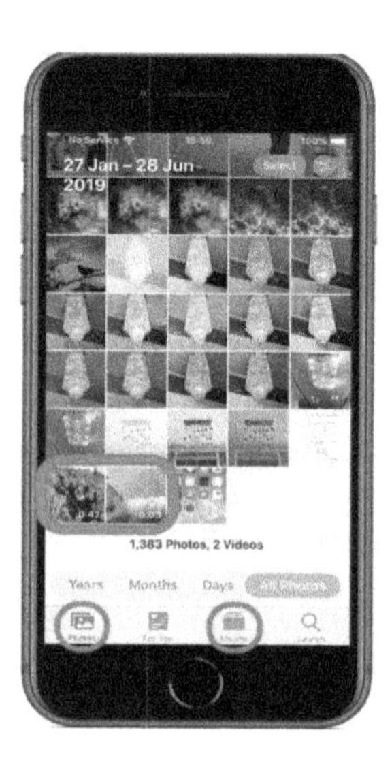

- Click on the Filter menu located at the end of your screen (it has a shape like a Venn diagram)

- Move through the available 9 filters to see how each would look on your video.
- Select your preferred filter and you would see a horizontal dial under the filter you selected.
- With your finger, slide the dial to adjust the level of intensity of that filter.
- Click on **Done** at the right bottom side of your screen to effect the filter on your video.

How to Use Lighting Mode Photo Effects in iOS 13

When you capture a picture while in portrait mode, the device makes use of the dual camera to create a depth-of-the-field effect which would allow you to create a photo that has a blurred background with a sharp subject. The iOS 13 introduced another feature called the

High-Key Light Mono. This is a white and black effect like the Stage Light Mono but rather than add a black background, you would get a white background.

- Go to the Photo app on your device.
- Click on a portrait photo from your library to select it.
- Confirmed that the image was captured in a portrait mode by checking for the portrait label usually at the left top corner of your screen.

- Then click on **Edit** at the right top corner to go into an editing mode.

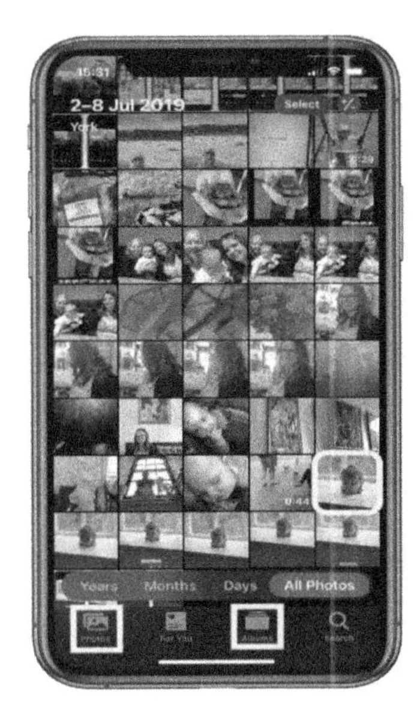

- Select the portrait icon in the tools at the bottom row then choose a lighting mode by sliding along the icons under the photos with your finger.
- After you have selected a lighting mode like the High-Key Light Mono effect, a slider would appear below it.
- Move your fingertip along the slider to rachet up or dial down the lighting effect intensity.
- Click on **Done** once you are satisfied with how the image looks.

How to Setup Vibration

- From the Home screen, go to **Settings.**

- Click on **Sounds & Haptics.**
- Toggle the switch next to **"Vibrate to Ring"** to enable or disable vibration when the silent mode is disabled.
- Toggle the switch next to **"Vibrate on Silent"** to enable or disable vibration when the silent mode is enabled.
- Return to the home screen.

How to Use the New Gestures for Copy, Cut, Paste, Redo and Undo

For most users, the iPhone has become the major way we communicate with people online as well as carry out other document features which is why it is important to have a good text management feature other than the "shake to undo" gesture in the old iOS.

With the iOS 13, Apple introduced the three-finger gesture to make it easy for typing. Once you get used to these features, you would enjoy communicating via your iPhone.

How to Redo and Undo

The **shake to undo** gesture has not been removed from iOS, however the three-finger swipe gesture is sure to override it as users get used to this new addition.

- To undo, swipe to the left with your 3 fingers on the screen.
- To Redo, swipe to the right with your 3 fingers on the screen.
- Another way to undo is by double-clicking on the screen with your three fingers.
- If you look at the top of your screen, you would see the badges for "Redo" or "Undo" to verify your action.

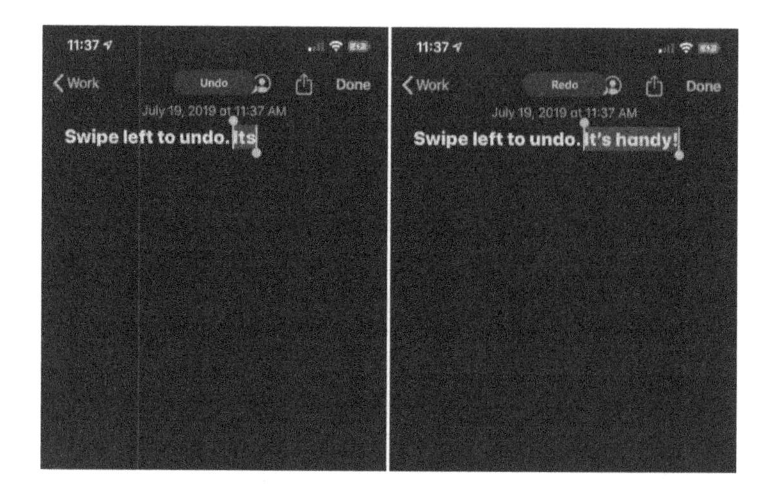

How to Copy, Cut and Paste

To best perform this feature, I would advise you use your two fingers and your thumb and this can be somewhat tricky if using a small screen.

- To copy, use your 3 fingers to pinch on the text and then un-pinch (expand) using your three fingers to Paste the copied text.
- Perform the copy gesture twice with your finger to cut out text. While the first gesture would copy the text, repeating it the second time would cut out the text.
- If you look at the top of your screen, you would see the badges for "Copy," "Cut," or "Paste" to verify your action.

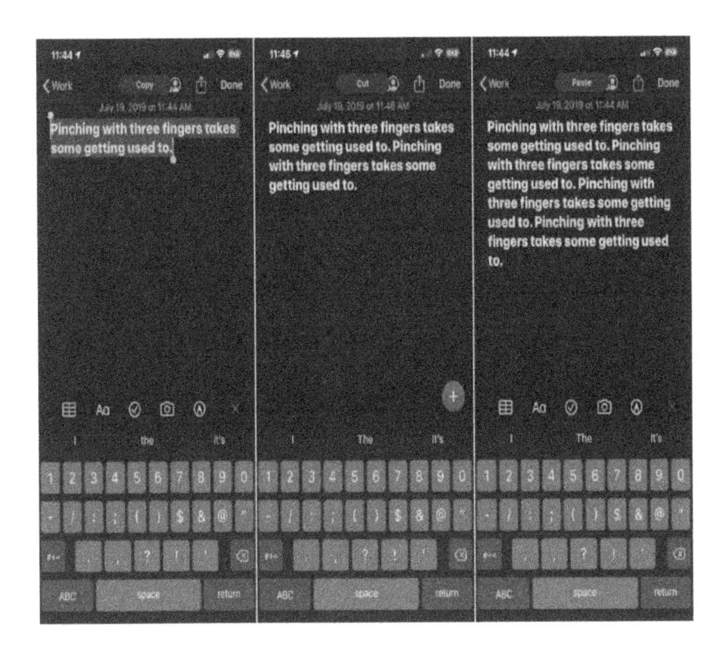

Cursor Movement on iOS 13

The way the cursor is moved has also changed with the iOS 13. To drag around a text entry, simply click on the entry cursor.

You no longer have to wait for some time before picking up text, once you touch it, you can move it immediately to where you want to place it.

Sadly, the little magnifying glass that pops up on the screen is no more and this can make it a little difficult to make a precise cursor placement as you would be unable to see the characters that you have picked.

How to Access the Shortcut Bar

If you think that the gestures are difficult to use, then you can make use of the new shortcut bar. On any part of the screen, including the keyboard area, click and hold on the screen with your 3 fingers for approximately a second to launch the shortcut bar. The bar would appear above the screen with buttons for **Cut, Undo, Copy, Redo and Paste.** You will continue to see the bar as you repeat command and then disappears as soon as you enter text or move the cursor.

List of New Keyboard Shortcuts

During the WWDC conference, Apple noted that they included about 30 new shortcuts in Safari which has been compiled below for your reference.

- For the default font size in Reader (Cmd + 0)
- Increase Reader text size (Cmd + +)
- Decrease Reader text size (Cmd + -)
- Actual size (Cmd + 0)
- Toggle downloads (Cmd + Alt)
- Open link in background (Cmd + tap)

- Open link in new tab (Cmd + Shift + tap)
- Open link in new window (Cmd + Alt + tap)
- New Private tab (Cmd + Shift + N)
- Close other tabs (Cmd + Alt + W)
- Save webpage (Cmd + S)
- Zoom out (Cmd + -)
- Email this page (Cmd + I)
- Use selection for Find (Cmd + E)
- Zoom in (Cmd + +)
- Focus Smart Search field (Cmd + Alt + F)
- Close web view in app (Cmd + W)
- Change focused element (Alt + tab)
- To download linked file (Alt + tap)
- Add link to your Reading List (Shift + tap)
- Paste without formatting content (Cmd + Shift + Alt + V)
- To toggle bookmarks (Cmd + Alt + 1)
- Navigate around screen (arrow keys)
- Open search result (Cmd + Return)

How to Set Screen Brightness

- From **Settings,** go to **Display & Brightness.**

- Under the **Brightness** option, click on the indicator and drag either to the left or to the right until you get your desired brightness.

- Click on the >Back sign.

- Click on **General** then **Accessibility.**

- Next, select "**Display Accommodations".**

- Beside the **Auto-Brightness,** slide the button to the left or right to either switch on or switch off this option.

How to Control Notification Options

- From Settings, go to **Notifications.**
- Click on **Show Preview** and set to **Always** to be able to preview notification on lock screen.
- To set this to only when the device is not locked, click on the option "**When Unlocked".**
- To disable notification preview, select "**Never".**
- Click on the Back arrow at the top left of the screen.

How to Control Notification for Specific Apps

- From the last step above, Click on the specific application.
- On the next screen, beside **Allow Notifications,** move the slide left or right to enable or disable.

How to Control Group Notification

- Scroll down the page and click on **Notification Grouping.**
- Select any of the 3 options as desired.
- Use the Back button to return.

How to Set Do Not Disturb

Your device can be put to silent mode for defined period. Even though your phone is in silent mode, you can set to receive notification from certain callers.

- Under **Settings,** click on **Do Not Disturb.**
- Toggle the switch next to "**Do Not Disturb**" to enable or disable this function.
- Toggle the switch next to "**Scheduled**" then follow instructions on your screen to set the period for the DND.
- Under **Silence** chose **Always** if you want your device to be silent permanently.
- Select "**While iPhone is locked**" if you want to limit this to only when the phone is locked.
- Scroll down and click on "**Allow Calls from**".

- Chose the best setting that meets your need to set the contacts that can reach you while on DND.
- Click on the back arrow at the top left of the screen.
- Scroll down to **Repeated Calls** and switch the button on or off as needed.
- Click on "**Activate**" under "**Do Not Disturb While Driving**".
- On the next screen, chose your preferred option.
- Click on the back button to return to the previous screen.
- Scroll down and select "**Auto Reply To**".
- On the next screen, select the contacts you wish to notify that **Do Not Disturb While Driving** is on.
- Go back to the previous screen.
- Scroll down and select **Auto Reply,** then follow the instructions on the screen to set your auto response message.

How to Turn PIN On or Off

- From **Settings,** click on **Phone.**

- At the bottom of the screen, click on **SIM PIN.**
- Turn the icon beside SIM PIN to the left or right to put off or on.
- Put in your PIN and click on **DONE**. The default PIN for all iPhone XR is 0000.

How to Change Device PIN

- From **Settings,** click on **Phone.**
- At the bottom of the screen, click on **SIM PIN.**
- To change PIN, click on **Change PIN**.
- Type in your current PIN and click DONE.
- On the next screen, type in the new 4-digit PIN and tap DONE.
- The next screen would require you to input the PIN again and click on DONE.

How to Unblock Your PIN

If you enter a wrong PIN 3 consecutive times, it would block the PIN temporarily. Follow the steps to unblock:

- On the home screen, click on **Unlock.**
- Put in the PUK and click on OK.
- Set a new 4-digit PIN and click **OK.**
- Input the PIN again and confirm.

How to Confirm Software Version

- From **Settings,** go to **General** and click on **About.**
- You would see your device version besides **Version** on the next screen.

How to Update Software

- From **Settings,** go to **General** and click on **Software Update.**
- If there is a new update it would show on the next screen.
- Then follow the screen instruction to update the software.

How to Control Flight Mode

- From the top right side of the screen, slide downwards.

- Tap the aero plane sign representing flight mode icon to turn off or on.

How to Choose Night Shift Settings

- From **Settings,** go to **Display & Brightness.**
- Click on **Night Shift.**
- Beside **Scheduled,** click on the indicator and follow the instruction on the screen to select specific period for the Night Shift.

How to Control Automatic Screen Activation

- From **Settings,** go to **Display & Brightness.**
- On the next screen, beside **Raise to Wake,** move the slide left or right to enable or disable.

How to enable Location Services/ GPS on your iPhone XR

- From the Home screen, choose the **Settings** option.

- Scroll towards the bottom of the page and click on **Privacy.**
- Then click on **Location Services.**
- Click on all the apps you would like to have access to your location data.
- Once selected, chose the option **While Using the App.**

How to Turn off location services on iPhone selectively

If there are any apps you would like to block from accessing your location, you can easily turn off location service for such apps by following the steps below.

1. Go to settings on the phone.
2. Move down to the **Privacy** option and then select **Location Services.**
3. You would see all the apps that have access and don't have access to your location. For the apps you wish to access your location information, find such apps, click on them and select **While Using the App.** For the apps you do not wish to access your location information, find such app, click on it and select **Never.** You can also use these steps

for the system services you wish not to grant access to your location information.

How to Turn off location services on iPhone completely

If you do not want any apps or systems on your iPhone to access your location information, follow the steps below to disable it:

- Go to **settings** on the phone.
- Move down to the **Privacy** option and then select **Location Services**
- To turn off the location service, all you need to do is toggle the button and then select **Turn off** to confirm the action. This would prevent all apps and system services from gaining access to your location data.

How to Use Music Player

- Click on the **Music Player** icon on the home screen.
- Click on **Playlist** then click on **New Playlist.**
- Tap the text box that has **Description,** type in the name for that playlist.

- Click on **Add Music.**
- Go to the category and click on the audio file you want to add.
- Select **Done** at the top of the screen.
- Select **Done** again.
- Go to the playlist and click on the music.
- Use the Volume key to control the volume.
- Click on the song title.
- Tap the right arrow to go to the next music or the left arrow to go to the previous music.
- Gently slide your finger up the screen.
- Click on shuffle to set it on or off.
- Click on Repeat to set it on or off. Here you can select the number of files you want repeated.

How to Navigate from the Notch

Both the sensors and the Face ID cameras are located in the notch found at the top of the device screen. With the notch, you are able to tell the difference between two important gestures which are the notification center and the control center.

1. **Steps to View Notification Center**

You can access the notification center by swiping down from the notch itself or from the left side of the notch.

2. **Steps to View Control Center**

From the right side of the notch, swipe down to view the control center.

Although the notch has occupied most of the space meant for the status bar, however, once you get into the Control center, you would be able to see all the status bar, this includes the percentage of your battery.

How to Use Swipe Typing

Another addition to the iOS 13 is the swipe typing feature in your device default keyboard. This feature allows you to type a word by swiping your fringe across the keys of your keyboard rather than tapping out each word. The keyboard world then sorts out the rest by deducing the right words you were trying and then inserting it into your message. Although it may take some time for you to get used to it, seeing that it is a new development, however once you get the hang of it, you would notice that it is actually faster than tapping each key.

Swipe typing is already enabled with the iOS 13 and so you would not need to take any action to turn it on. For instance, say you want to type “call", all you need to do is to tap on the “c" key with your keyboard the day your finger over the “a", “l" and “l" keys in this order. The keyboard world automatically predicts the words you wish to type.

How to Disable Swipe Typing

While the swipe typing is enabled by default, you can disable it if you do not find it useful or easy.

- Go to the settings app.
- Click on **General.**
- Then select **Keyboards.**
- Navigate to **"Slide to Type"** and move the switch to the left to disable the option.

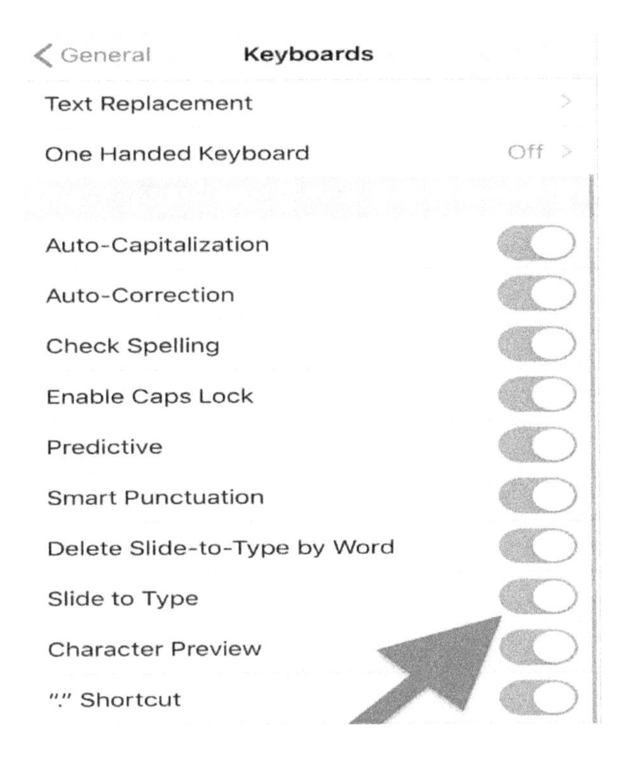

Another way to do this is by using the keyboard itself.

- Long press on the keyboard switcher.
- Then click on **Keyboard settings.**
- Depending on your keyboard setup, you would see the switcher as a Globe icon or an Emoji icon.
- Navigate to **"Slide to Type"** and move the switch to the left to disable the option.

Note: If you have only one keyboard activated on your device, you would be unable to use this option.

How to Disable/ Enable Haptic Touch and 3D in iOS 13

The 3D touch was introduced in 2015 with the iPhone 6s to help you navigate, view and control features and apps in the device. The 3D allows you to access preview notifications, launch Quick Actions and operate actions from the control center along with other features. While some people considered it helpful, some saw it as annoying and would like to disable it. iOS 13 replaces the 3D touch with the Haptic Touch. Even at that, some people are not fans of the Haptic Touch either. If you belong to this class, you can disable the function through the steps below:

- Go to the settings app.
- Click on **General.**
- Select **Accessibility.**

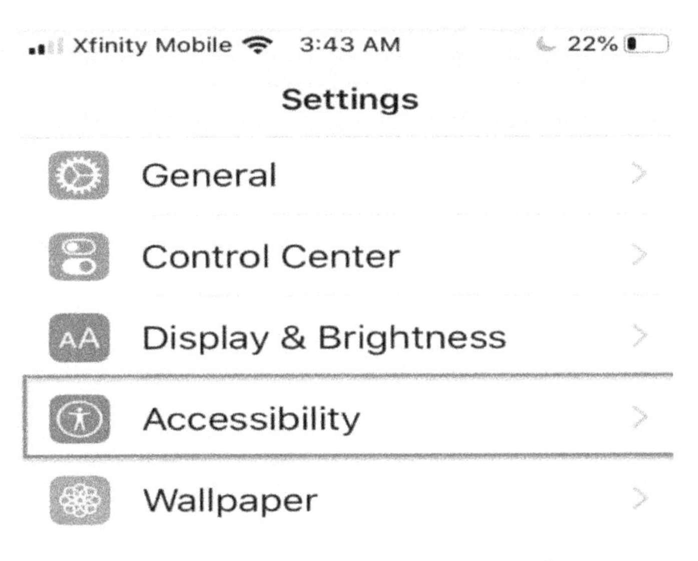

- Then click on **3D and Haptic Touch.**

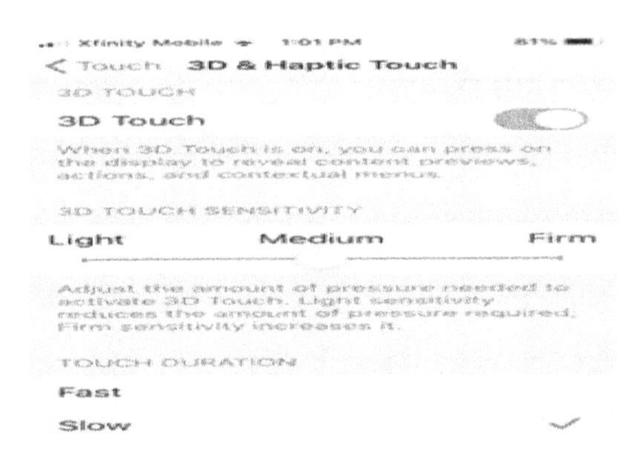

- Move the switch to the left to disable **3D Touch.**

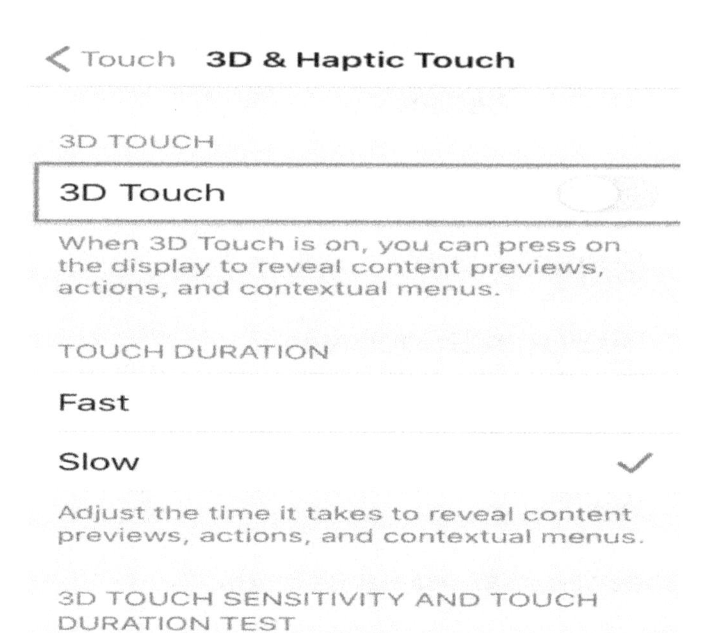

Go through the steps to enable the option whenever you wish to

How to Remove Location Details from your Photos in iOS 13

The **iOS 13** has made it possible to remove location details from your photos or choose the persons you would like to have access to this information. You can remove location from your videos, photos, movies or multiple images that you intend to send via messages, Facebook, Mail, and so on with the steps below:

- Capture your photos in the normal way with your camera app.
- Go to the section where the photo was saved.
- If sharing a single video or photo, click on it to open, then click on the **Share** button.
- If sharing several videos and photos, click on **Select** in the section view, then click on all the items you wish to share and then click on the **Share** button.
- On the next screen to share, you would notice a new button for **Options.**
- Click on the **Options** and disable the **Location** on the next screen.
- Then select any media you like to send your photo through.

Note: You would have to set this feature each time you want to share a video or picture. The Location details can only be disabled from the iPhones photo app so it is important that you always share your contents from the app directly. The photos and videos on your phone would contain to have the location details as the feature only affects contents that you want to share to third party.

How to keep Track of documents

On your iPhone XR, you can access folders and files stored on your iCloud Drive and any other cloud storage services. You can also access and restore folders and files deleted from your device within the last 30 days. There are 3 subsections in the Browse tab and they are:

1. **Locations:** To view files saved in iCloud, simply click on **iCloud Drive.** To view files recently deleted from your device, click on **Recently Deleted.**

   To add an external cloud storage service, you need to first install the app from the App store (Google Drive, Dropbox etc.), then click **Edit** at the right top corner of your device screen to activate it. Once done, click on **Done.**

   Other available options for folders and files are:

   - To view the content of a folder, click on the folder.

- To Copy, Rename, Duplicate, Delete, Move, Tag, Share or Get info of a folder or file, simply press the folder or file for some seconds.
- To download items with the cloud and arrow icon, tap on them.
- To annotate a file, simply click on the pencil tip icon at the right upper side of the screen. It is important to know that this feature is only available for select image file formats and PDF.

2. **Favorites:** To add folders to the Favorite section, press the folder for some seconds until a menu pop up, then select **Favorite** from the menu. Currently, you can only do this from the iCloud Device, and only for folders, no single files.
3. **Tags:** when using macros Finder tags, you will see them in the Tags section. Alternatively, press a file for some seconds to tag such file. Then click **"A Tag Here"** to see all the files that have that tag.

How to Move Between Apps

Switching Apps can be tricky without a home button, but the below steps will make it as seamless as possible.

- To open the App Switcher, swipe up from the bottom of the screen and then wait for a second.
- Release your finger once the app thumbnails appear.
- To flip through the open apps, swipe either left or right and select the app you want.

How to Force Close Apps in the iPhone XR

You do this mostly when an app isn't responding.

- Simply swiping up from the bottom of the screen would show the app switcher. This would display all the open apps in card-like views.
- For iOS 12 users, to force close the app, locate the app from the app switcher and swipe up to close the app.
- For users still on iOS 11, press the app you wish to close for some seconds until you see the red button marked with the minus sign at the top of each app card.

- Tap on the minus button for each of the app you wish to close.
- I would advise you upgrade to iOS 12 to enjoy better features on your iPhone XR.
- To go through apps used in the past, swipe horizontally at the bottom of your home screen.

How to Arrange Home Screen Icons

Follow the steps below to arrange the homes screen icon on your iPhone XR.

- Press and hold any icon until all the icons begin to wiggle.
- Drag the icons into your desired position.
- Tap either the **Done** button at the right upper side of the screen or swipe up to exit the wiggle mode.

Complete iPhone XR Reset Guide: How to perform a soft, hard, factory reset or master reset on the iPhone XR

Most minor issues that occur with the iPhone XR can be resolved by restarting the device or doing a soft reset. If the soft reset fails to solve the problem, then you can carry out other resets like the hard reset and master reset. Here, you would learn how to use each of the available reset methods.

How to Restart your/ Soft Reset iPhone

This is by far the commonest solutions to many problems you may encounter on the iPhone XR. It helps to remove minor glitches that affects apps or iOS as well as gives your device a new start. This option doesn't delete any data from your phone so you have your contents intact once the phone comes up. You have two ways to restart your device.

Method 1:

- Hold both side and Volume Down (or Volume Up) at the same time until the slider comes up on the screen.
- Move the slider to the right for the phone to shut down.
- Press the **Side** button until the Apple logo shows on the screen.
- Your iPhone will reboot.

Method 2:

- Go to **Settings** then **General.** Click on **Shut Down.**
- This would automatically shut down the device.
- Wait for some seconds then Hold the **Side** button to start the phone.

How to Hard Reset/ Force Restart an iPhone XR

There are some cases when you would need to force-restart your phone. These are mostly when the screen is frozen and can't be turned off, or the screen is unresponsive. Just like the soft reset, this will not wipe the data on your device. It is important to confirm that the battery isn't the cause of the issue before you begin to fore-restart.

Follow the steps below to force-restart:

- Press the **Volume Up** and quickly release.
- Press the **Volume Down** and quickly release.
- Hold down the Side button until the screen goes blank and then release the button and allow the phone to come on.

How to Factory Reset your iPhone XR (Master Reset)

A factory reset would erase every data stored on your iPhone XR and return the device back to its original form from the stores. Every single data from settings to personal data saved on the phone will be deleted. It is important you create a backup before you go through this process. You can either backup to iCloud or to iTunes. Once you have successfully backed up your data, please follow the steps below to wipe your phone.

- From the **Home** screen, click on **Settings.**
- Click on **General**.
- Select **Reset.**

- Chose the option to **Erase All Content and Settings.**
- When asked, enter your passcode to proceed.
- Click **Erase iPhone** to approve the action.

Depending on the volume of data on your phone, it may take some time for the factory reset to be completed.

Once the reset is done, you may choose to setup with the **iOS Setup Assistant/Wizard** where you can choose to restore data from a previous iOS or proceed to set the device as a fresh one.

How to Use iTunes to Restore the iPhone XR to factory defaults

Another alternative to reset your phone is by using iTunes. To do this, you need a Computer either Mac or Windows that has the most current version of the iOS as well as have installed the iTunes software. Factory reset is advisable as a better solution to major issues that come up from software that wasn't solved by the soft or force restart. Although you would lose data, however, you get more problems fixed including software glitches and bugs.

Follow the guide below once all is set:

- Use the Lightning Cable or USB to connect your device to the computer.
- Open the iTunes app on the computer and allow it to recognize your device.
- Look for and click on your device from the available devices shown in iTunes.
- If needed, chose to back up your phone data to iTunes or iCloud on the computer.
- Once done, tap the **Restore** button to reset your iPhone XR.
- A prompt would pop-up on the screen, click **Restore** to approve your action.
- Allow iTunes to download and install the new software for your device.

How to Choose Network Mode

- From **Settings**, go to **Mobile Data**.
- Select **Mobile Data Options> Enable 4g.**
- To stop using 4g, choose **Off.**
- This option would make your device to automatically switch to either 2g or 3g depending on available coverage.

- Click on **Voice & Data** if you want to use 4g for both mobile data and voice calls.
- Note: To get fast and better connection, use 4g for calls via the mobile network.
- Click on **Data Only** to use for only mobile data.
- Done.

How to set a reminder on iPhone XR

Follow the steps below to create a reminder.

1. Open the **Reminder app** on your iPhone XR device.

2. At the top right corner of the screen, click the plus button to create a new reminder or a list.

3. To create a list, tap **List** and tap **Reminder** to create a new reminder.

4. For reminder, enter the exact reminder content.

5. In the content box space, you have two choices.

   First option: remind me on a day. With this option, please set the Alarm and Repeat options – Every day, Every week, every month, never etc.,

Second option: Remind me at a location. For this, turn your location on, then set the location you will receive when you arrive or leave.

6. Choose the priority level for the reminder, you can also add notes if needed

7. Chose **Done** to complete the process.

How to set a Recurring Reminder on your iPhone XR

To create a recurring reminder on your device, follow the steps below:

1. Go to the Reminders app on your device.

2. Type your content on the space for reminder content.

3. Click on the info button beside the new reminder set.

4. Select the option to **"Remind me on a day".**

5. Set the time you want the reminder.

6. Select **Repeat** and then Custom.

7. Set your frequency to Repeat, Weekly, Daily, Monthly or Yearly.

8. Once done, go to **End Repeat** and select date you want the reminder to stop.

How to get Battery Percentage on iPhone XR

The iPhone XR does not give room to see the battery percentage of your iPhone always, but you can always get a peek to see your battery percentage. To do this, place your finger at the top-right corner of the iPhone XR display and swipe down to be able to access the Control Center. Once the control center is open, you would see the battery percentage at the top right corner of the page.

How to take a Screenshot

Without the home button, taking a snapshot may seem tricky; however, follow these steps to help you take the best shots possible.

- Press both the side and the Volume Up button simultaneously to take a screenshot.

- The photo from the screenshot would be saved automatically in the Photos app, under the **Screenshots** album. Screenshots help you to note down problems you wish to seek help for later.
- To edit the photo, go to the photo and tap the thumbnail at the left bottom corner of your iPhone.
- To view the screenshots in iOS 11, go to **Photos** click on **Albums** then **Camera Roll/ Screenshots**. To do same in iOS 12, go to **Photos,** then **Albums,** go to **Media Types** and select **Screenshots.**

Chapter 3: The New Reminder App and Apple Map

While the iOS 13 brought about entirely new features, it also modified the old ones with new additions. The default Apple reminders app was lacking in several features that you would find in other third-party to-do apps. Now the iOS 13 has modified the reminders app to include all the features that users would like. The steps below would guide you on how to use the new reminders app.

How to Use the New Reminders app

The iOS 13 brought a new built for the iOS reminders app. When you launch the reminders app, you can view the total reminders you have at the moment, the ones that are due today and the numbers that can be seen in each list. To add a new task to the list,

- Click on **All.**
- Then click the "+" button located underneath each category to add a new task

You can also add a reminder time or date, change the category for a task or set to be reminded of a task in a specified location by clicking on the 'I' icon in blue icon once you have tapped on the desired tasks to launch the options.

How to Create a Reminder

- Launch the reminders app.
- Then click on **reminders** under **My List** heading
- Then select **New Reminders** at the bottom of your screen, at the left side.
- Fill in your details for the reminder.
- Then click **Return** on your keyboard to confirm your first reminder.

How to Add Location, Time or a Connected person

It's one time to add a reminder, it's another thing to have do go through your to-do to be reminded of what you intend to do in the day. Now you can add a time or location to your reminder so that your phone can prod you at the right time or place. After you have set your reminder, click on the blue "i" icon that is located at the right side of your task to access the task options.

- To be reminded at a specific time, enable the option for ***Remind me on a day*** *and then fill in details in the option for* **Alarm or Reminds me at a *time.***

- *For recurring tasks, you can set the option to repeat.*

- *To add location, you should go for the option of* ***Remind me at a* location** and then select your desired location

- If your task has to do with someone, then you can set the reminder to alert you about the task whenever you message the contact. To do this, click on ***Remind me when messaging*** *and then choose your preferred contact.*

Get Siri to Remind You

With your virtual assistant, you do not always have to launch the reminders app and begin inputting details to remove an event or activity. You can say to Siri, "remind me to," followed by the contents of what you would like to be reminded on. You can also ask Siri to remind you at a specific place or time.

How to Add SubTasks

For complex projects or tasks, you can add sub tasks or you can even create a multi entry list for your shopping. If you have a reminder that needs to have the subtasks set,

- Go to the task from your reminder app.
- Click on the blue "i" icon to launch the options.
- Then navigate down on your screen to the option for **Subtasks.**
- Click on it and then tap **Add Reminder** to include a subtask.
- Feel free to add as many subtasks as you like.
- When you are done, you would find your subtask under the reminder or the main task.
- You can complete the subtasks separately from the parent task.
- You can also click on the "i" icon to add its own time, separate location or contacts for each individual subtask.

How to Use Today Notification Feature

You may have noticed that the reminders app has a new home page which is self explanatory. However, there is one important feature that is not revealed. By default, the reminders app would notify you on the tasks you have for each day. But you can change when you want the notification to happen or if you want to be notified at all. To effect this change,

- Go to the settings app.
- Click on **Reminders.**
- On the next screen, you can completely turn off notifications or change the timing door the option of **Today Notification.**
- In this same screen, you can modify your tasks' default list.

How to Create a List

This new modifications in the reminder app can get us carried away with filling every little detail from tax notifications, to birthdays to grabbing milk on your way from work. The good thing is that you can organize your tasks into lists to declutter your reminders home page. The steps below would show you how to create a list.

- Launch the reminders app
- Then click on **Add list** found at the right bottom of the screen.
- Select from the varieties of logos and colors to help you tell the different lists at a glance.
- Once satisfied with your list, click on **Done** at the right top side of your screen.
- To add reminders or tasks to the list, click on the list from under the ***My Lists*** subheading and create them or move existing tasks to the list
- For existing task, open the task, then click on the "i" icon to access the options.
- Go down and click on **List** then choose from your new list.

Note: if you are not able to easily find a task, use the search bar option in the reminders app to find the task.

How to Add a List to a Group

Now you have created a list and moved the tasks to the list. However, there is more. You can add lists in same categories to a general group. So, for instance, you have a list containing anniversaries and another one containing birthdays, you can group both of them into an 'important date' group to keep your home page looking appealing.

Follow the easy steps below to create your group

- Go to the homepage for the reminders app
- At the right top corner of your screen, click on **Edit.**
- Then click on **Add Group** at the bottom left of your screen.
- Input your preferred group name and choose all the lists you want to add to the group.
- Then tap **Done.**
- To modify the lists in each group, click on **Edit** again.

- Then click on the "i" icon next to the group.
- Then remove or add lists using the ***Include*** option.

How to Use Favorites in the Apple Maps

I know that Apple map has been available for some time now but not so many like using it. Good news is that the IOS 13 has brought an improvement to the Apple map to include more beaches, roads, building and other details you may be interested in. Apart from these listed ones, there are also some other cool features that were just added like being able to add a location to your list of Favorites. You can also arrange the saved locations in your own personally customized collections. Follow the steps below to add a favorite on the map.

- Search for a location or tap on a location.
- Scroll down to the bottom and click on **Add to Favorites.**
- You can always access your favorites list on your main page.

To add a particular location to your customized collection,

- Drag up from the Apple maps main page.
- Then click on **My Places.**
- Select **Add a Place.**
- On the next screen, you can now add any location that you recently viewed to your collection or search through your search bar for the location.

To begin a new collection,

- Navigate back to your apple maps main page.
- Swipe from the bottom of the screen upwards.
- Then click on **New Collections** to make a new list.

How to Use the Look Around Feature in Apple Maps

Look around is Apple's version of the Streetview from Google as it allows you to preview a location before you visit. Follow the steps below on how to use it.

- Type in a location on your Apple maps.

- Then select it by pressing long on the map.

- If the location supports Look around you would find a **look around** image on the location.

- Click on it to move down to Street level and drag to navigate around.

- While on this view, you can also see facts about the place or even add it to your favorites list but swiping up from the bottom of the screen.

At the moment, Apple hasn't covered the whole locations in the USA but they have promised it do this by end of 2019 and also follow suite for other countries.

CHAPTER 4: Calls and Contacts

How to Make Calls and Perform Other Features on Your iPhone XR.

This section would give you a detailed guide to contacts and call management in your device.

How to Call a Number

- Tap the Phone icon at the left.
- Click on Keypads to show the keypads.
- Input the number you want to call then press the call icon.
- Tap the end call button at the bottom of the screen once done.

How to Answer Call

- Tap any of the volume keys to silence the call notification when a call comes in.
- If the screen lock is active, slide right to answer the call.
- Click on Accept, if there is no screen lock.
- Tap the end call button at the bottom of the screen once done.

How to Control Call Waiting

- From **Settings,** click on **Phone** then **Call Waiting.**
- Move the icon beside it to the left or right to enable or disable call waiting.

How to Call Voicemail

- Click on the phone icon at the left of the home screen.
- Select **Voicemail** at the bottom right corner of the screen.
- Click on **Call Voicemail** in the middle of the screen and listen for the instructions.
- Tap the end call button at the bottom of the screen once done.

How to Control Call Announcement

Your device can be set to read out the caller's name when there is an incoming call. The contact has to be saved in your address book for this to work.

- From **Settings,** go to **Phone** then **Announce Call.**
- Select **Always** if you want this feature when silent mode is off.

- Choose **Headphones & Car** to activate when your device is connected to a car or a headset.
- The **Headphones Only** option would be for when the device is connected to only headset.
- Select **Never** if you do not wish to turn off this feature.

How to Add, Edit, and Delete Contacts on iPhone XR

Follow the steps below to add, edit and delete contacts on your new device.

How to Add Contacts

- At the **Home** screen, select **Extras.**
- Click on **Contacts.**
- Then select the **Add Contact** icon at the right upper side of your screen.
- Enter the details of your contact including the name, phone number, address, etc.
- Once done inputting the details, tap **Done** and your new contact has been saved.

How to Save Your Voicemail Number

- Once you insert your SIM into your new device, it automatically saves your voicemail number.

How to Merge Similar Contacts

- At the **Home** screen, select **Extras.**
- Click on **Contacts.**
- Click on the contact you want to merge and click on **Edit.**
- At the bottom of the screen, select **Link Contact….**
- Choose the other contact you want to link.
- Click on **Link** at the top right side of the screen.

How to Copy Contact from Social Media and Email Accounts

- From Settings, go to **Accounts and Password.**
- Click on the account, e.g. Gmail.
- Switch on the option beside **Contacts.**

How to Create New Contacts from Messages On iPhone XR?

- Go to the Messages app.
- Click on the conversation with the sender whose contact you want to add.
- Above the conversation, you would see their phone number.
- Click on the phone number.
- This would show 3 buttons on the screen.
- Click on the **Info** option.
- You would see the number again at the top of the screen, click on it.
- Then click **Create New Contact.**
- Input their name and other details you have on them.
- At the top right hand of the screen, click on **Done**.

How to Add a Caller to your Contact

- On your call log, click on a phone number.
- You would see options to **Message, Call, Create New Contact or Add to Existing Contact**.
- Select **Create New Contact.**

- Enter the caller's name and other information you have.
- At the top right hand of the screen, click on **Done.**

How to Add a contact after dialing the number with the keypad

- Manually type in the numbers on the phone app using the number keys.
- Click on the (+) sign at the left side of the number.
- Click on **Create New Contact.**
- Enter the caller's name and other information you have.
- Or click on **Add to Existing Contact**.
- Find the contact name you want to add the contact to and click on the name.
- At the top right hand of the screen, click on **Done.**

How to Import Contacts

The iPhone XR allows you to import or move your contacts from your phone to the SIM card or SD card for either safekeeping or backup. See the steps below:

- From the Home screen, click on **Settings.**
- Select **Contacts.**
- Chose the option to "**Import SIM contacts".**
- Chose the account you wish to import the contacts into.
- Allow the phone to completely import the contacts to your preferred account or device.

How to Delete contacts

When you remove unwanted contacts from your device, it makes more space available in your internal memory. Follow the steps below.

- From the Home screen, tap on **Phone** to access the phone app.
- Select **Contacts.**
- Click on the contact you want to remove.
- You would see some options, select **Edit.**
- Move down to the bottom of your screen and click on **Delete Contact.**
- You would see a popup next to confirm your action. Click on **Delete Contact** again.
- The deleted contact would disappear from the available Contacts.

How to Manage calls on your iPhone XR

Here, we would talk about how to block calls, set or cancel call forwarding, manage caller ID as well as call logs on your device.

How to Block Spam Calls on iOS 13

The new "silence unknown callers" feature is another addition to the iOS 13. With this feature, you can block spam calls without the need to block each one separately. This way, if you find out that the call is not actually spam, you can just go to your voicemails to check for the call and call back anyone that you need to contact. Follow the steps highlighted below to activate the new call blocker:

- Go to your settings app.
- Click on **Phone.**
- Then move the switch beside ***Silence Unknown Callers*** to the right to enable it.

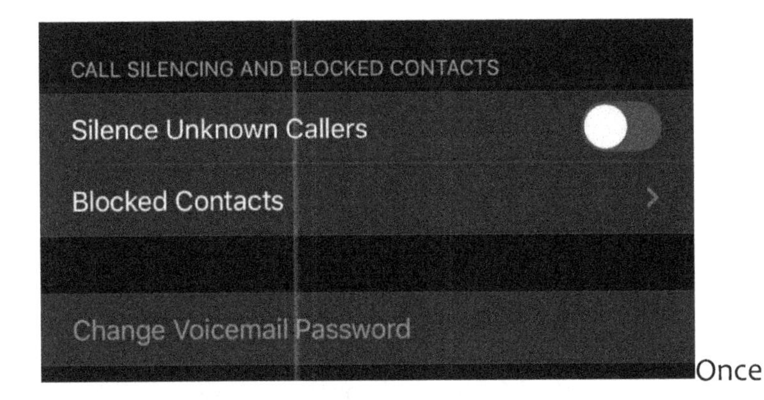

Once this feature is active, unknown callers would be sent straight to voicemail by default and won't have to worry about robocalls, spam calls and other distractions would no longer bother you.

How to Block Calls on the iPhone XR

- Go to **Settings** from the Home screen.
- Click on **Do Not Disturb.** (The DND feature on the iPhone XR allows you determine how you want your device to process incoming calls. The following options are available under DND.
    1. **Do Not Disturb** option – tap on this option to enable or disable the DND feature manually on the device.

2. Scheduled – To schedule a time for DND to be activate, just tap the tap the time and set the start to end time.
3. **Allow Calls From –** Use this option only when you want to receive calls from specific people. Select the people and allow calls from them.
4. **Repeated Calls** – This option allows a call to come through once the call is repeated within 3 minutes of the first call.

How to Block Specific Numbers/Contacts on Your iPhone XR

- Click on the **Phone** icon on the Home Screen.
- Tap **Recent** or **Contacts.**
- Select the specific contact(s) or number(s) you desire to block.
- If accessing through **Recent** option, tap the **(i)** icon next to the number.
- Click on **Block This Caller** at the bottom of the screen.
- Click on the **Block Contact** option to confirm your action.

- Blocked contacts or numbers would be unable to reach you.

How to Unblock Calls or Contacts on your iPhone XR

- Go to **Settings** from Home.
- Select **Phone-> Call Blocking & Identification** then click on **Edit.**
- Click on the **minus (-) sign** next to the contact or number you want to unblock.

How to Use and Manage Call Forwarding on your iPhone XR

With the Call Forwarding unconditional (CFU) feature in the iPhone XR, calls can be forwarded to a separate phone number without the main device ringing. This is most useful when you do not wish to turn off ringer or disregard a call but also do not want to be distracted by such calls. To enable this feature, follow the steps below:

- Go to **Settings** from Home.
- Click on **Phone** then **Call Forwarding.**
- Select the **Forward to** option.

- Input the number you want to forward such calls to.
- You can set the calls to be forwarded to voicemail.

Apart from CFU, Call Forwarding Conditional (CFC) allows you to forward incoming calls to a different number if the call goes unanswered on your number. To enable this feature, you need to have the short codes for call forwarding then set the options to your preference. For data on short codes, reach out to your carrier.

How to Cancel Call Forwarding on your iPhone XR

To cancel,

- Go to **Settings** then **Phone**
- Click on **Call Forwarding.**
- Move the slider to switch off the feature.

How to Manage Caller ID Settings and Call Logs on your iPhone XR

You can decide to hide your caller ID when calling certain numbers. Follow the steps below to activate this.

- Go to **Setting** on the Home screen.

- Click on **Phone** then click on **Show My Caller ID.**
- Click on the switch next to **Show My Caller ID** to either enable or disable the option.

When you disable the feature, the called party will not see your caller ID. This is usually for security or privacy reasons.

How to View and Reset Call Logs on your iPhone XR

For every call you make on your device, there is a log saved on the phone app. To view or manage the call log data, follow the steps below:

- On the Home screen, Click on **Phone** to go to the phone app.
- Click on **Recent** then click on **All.**
- Tap on the call log you wish to extract information from.

How to Reset Call Logs

- Go to **Phones,** then click on **Recent>All>Edit.**
- Click on the **minus (-) sign** to delete calls individually.

- To delete the whole call log once, simply tap **Clear** then chose the **Clear All Recent** option.

CHAPTER 5: Messages and Emails

How to Set up your Device for iMessaging

- From Settings, go to Messages.
- Enable iMessages by moving the slide to the right.

How to Set a Profile picture and Name in iMessages

With this feature, you can now set a screen name and a profile image on your iMessage that you would share with your selected contacts. So, when next you text another iPhone user, they would not need to save your contact details before they can know who is texting them. Follow the easy steps below to set it up.

- Open the messages app.
- Click on the 3 dots (…) at the right upper corner of your screen.
- Then click on **Edit Name and Photo.**
- On the next screen, you can choose a profile picture and also input your desired last and first name.

- You can choose to use your personal Memoji as your profile picture or choose from available Animoji.

- Then, set if you would like to share this detail with any of the options on the screen: with *Anyone*, *Contacts Only*, or to *Always Ask* if the details is to be shared.

How to Customize Your Memoji and Animoji

With the iOS 13, you no longer need to have an iPhone with TrueDepth to be able to create a Memoji. Although you still need a TrueDepth camera to have the Animoji or Memoji follow your movements, however, owners of the 6S or newer can now create their own Memoji, thanks to the iOS 13. To do this,

- Go to iMessage.

- Tap on a conversation to launch it.

- Then click on the Memoji icon, then tap the "+" button.

How to Create and Use Animoji or Memoji

The iPhone iOS 13 has made it possible for every device that has the iOS 13 to be able to access the Animoji or Memoji as you do not need to have an iPhone with TrueDepth selfie camera to be able to use this feature. So, you can now create a cartoon version of your loved ones or of self.

The steps below would show you how to create a Memoji

- go to name and profile picture settings.
- Click on the circle for pictures close to the name field.
- Then click on the "+" sign to make your own Memoji.

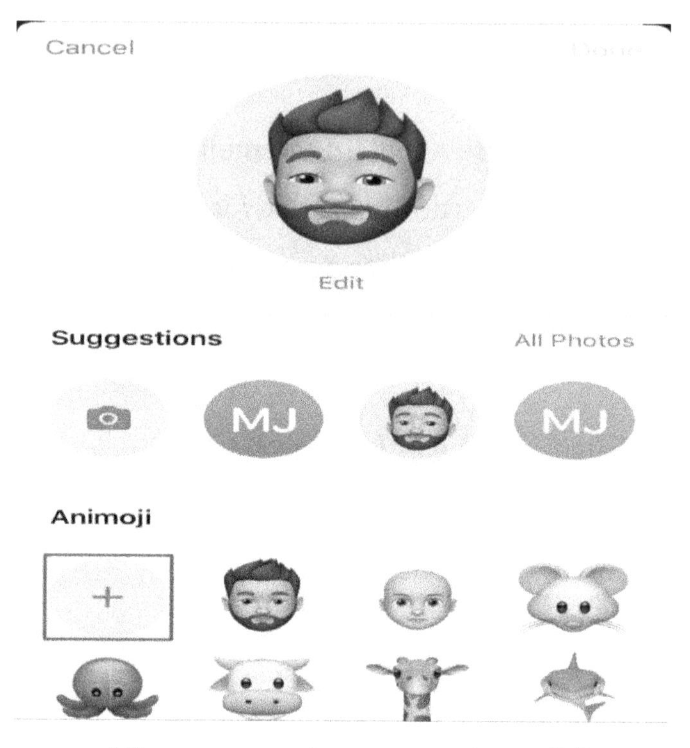

- After you must have created on, click on it to choose a pose for your Memoji and to also use it as your profile picture.

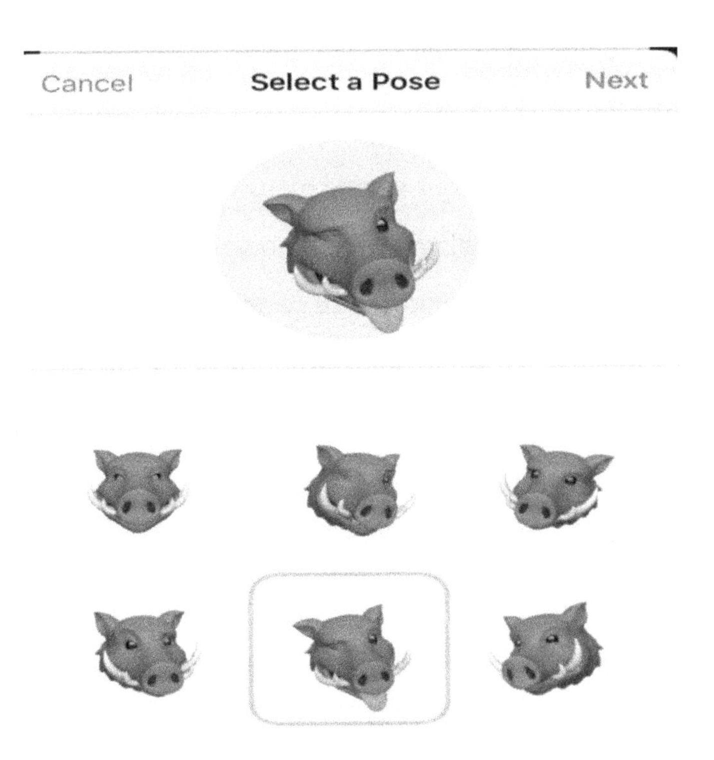

If you would rather use something else other than a picture of yourself, you can make use of the Animoji. In the Animoji menu selection, you would find several options to choose from including a shark, mouse and even a skull. Then pick a pose in a similar way with the Memoji.

After selecting an Animoji or Memoji, you have to scale it, place it to fit the circle then select a background color to finish the setup.

How to Set View for your Profile Picture and Name in iMessages

It is quite cool to have your profile picture and name set especially when messaging a friend, however you may not want an unknown person to have access to your real names or profile picture. Thankfully, there is a setting to limit who has access to what.

- Go to the settings for **Share Automatically,** then select from the 3 options available for sharing your name and profile picture.
- Go for **Contacts Only** if you would rather share this with only people in your contact list.
- The **Anyone** option means that everyone and anyone can access this information
- **Always Ask** would give you the option to personally choose people to share with. Each time you receive a message in your phone and you open it, you would receive a small pop-up at the top of your screen asking if you would like to share your details with the sender. Click on **Share** to send your details across or click on **"X"** to refuse and shut down the message.

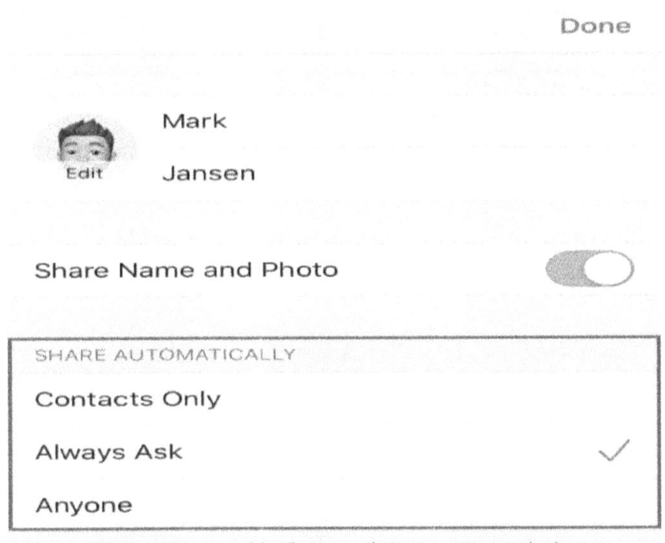

How to Compose and Send iMessage

- From the Message icon, click on the new message option at the top right of the screen.
- Under the "To" field, type in first few letters of the receiver's name.
- Select the receiver from the drop down.
- You would see iMessage in the composition box only if the receiver can receive iMessage.
- Click on the "Text Input Field" and type in your message.
- Click on the send button beside the composed message.

- You would be able to send video clips, pictures, audios and other effects in your iMessage.

How to Set up your Device for SMS

- Your device is automatically set up for SMS once you put in your SIM.

How to Compose and Send SMS

- From the Message icon, click on the new message option at the top right of the screen.
- Under the "To" field, type in first few letters of the receiver's name.
- Select the receiver from the drop down.
- Click on the "Text Input Field" and type in your message.
- Click on the send button beside the composed message.

How to Set up Your Device for MMS

- From **Settings**, go to **Messages.**
- Enable **MMS Messaging** by moving the slide to the right.

How to Block Spam, Contacts and Unknown Senders in iOS 13 Mail App

This feature would set the incoming emails from the blocked senders to go directly to the thrash folder. You are not really blocking the senders as you can always view the messages in the thrash folder. This option is better than totally blocking the contact. Currently, Apple has grouped all spammers and contacts that are blocked into a single folder. So, if in the past you blocked some phone numbers in Messages, FaceTime or Phone, you would see them along with the blocked senders for mails.

The first step is to select your block settings. This has to do with setting what you want the mail app to do with the blocked contacts.

- Go to the settings app.
- Click on **Mail.**
- Then navigate to **Threading.**
- Click on **Blocked Sender** and you would see 3 options

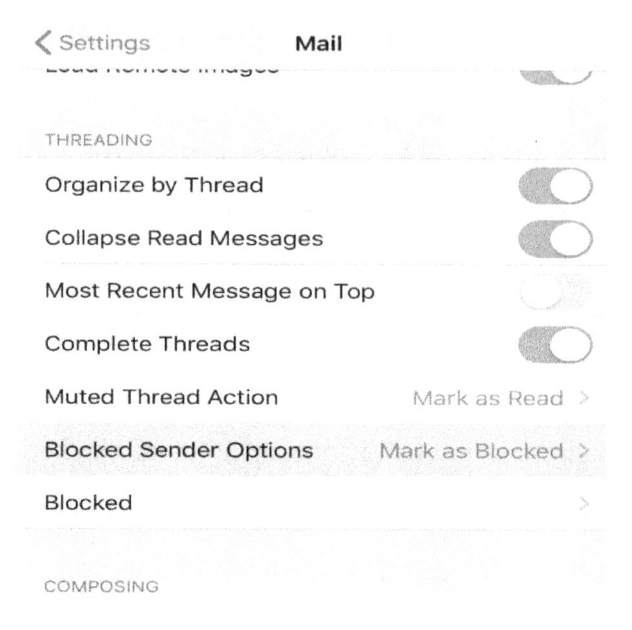

- If you choose **None,** it would disable email blocking.
- **"Mark as Blocked, Leave in Inbox"** means that the emails would come to your inbox but you would not be notified like the other emails.
- **Move to Trash,** would move emails from blocked contacts to your thrash folder. You can then set to manually empty the folder or have it deleted automatically.

Note: this feature would apply to all the accounts you have in your mail app including Outlook, Gmail, Yahoo etc.

How to Block a Sender from Received Emails

When you receive an email from someone you do not know or do not want to hear from again, you can block the person from the email.

- Click on the contact fields at the top of the email to show you all the parties in the send list.
- Then click on the email address you want to block.
- The next screen would show you an expanded menu option.
- Click on **Block this contact.**
- Click on it again in the prompt to confirm your action.
- The contact is blocked!
- All emails from the sender whether present or past would have a blocked hand icon close to the date in the header for email. You would also see a notification at the top of the email reading, **"This message is from a sender in your blocked list."**

How to Unblock a Sender from Received Email

If you change your mind about a sender, simply go back to the sender's contact details in the email and click on "Unblock this Contact." It may take some seconds for the hand icon to disappear but the sender would be unblocked instantly.

How to Block a Contact from Email Settings

In a case you do not have any email from the person you want to block, you can go to settings to stop them from sending any more messages.

- From the settings app, click on **Mail.**
- Then scroll down and click on **Blocked.**
- You would see all the email addresses and phone numbers that you have ever blocked in this screen.
- At the bottom of your screen, click on **"Add New."**
- Then select the contacts you want to block once prompted.

- You would then see the numbers and email addresses in the blocked list.

Note: if you do not have the contact of the person saved on your phone, you would need to follow the first option to block them directly from inbox. This option only works with saved contacts.

How to Unblock a Contact from Settings

You can unblock all categories of blocked people from your settings since every single contacts or senders blocked get to share on the settings menu. To do this, follow the steps above to access the list of your blocked contacts.

- The short-swipe from the left on the email address or phone number you want to want to unblock.
- Then click on **Unblock.**
- Another way is to long swipe on the contact to automatically unblock it.
- Or you can click on **Edit** at the right top of your screen, click on the red "-" minus button beside

the email address or phone number you want to unblock, then click on **Unblock.**

How to Compose and Send SMS

- From the Message icon, click on the new message option at the top right of the screen.
- Under the "To" field, type in the first few letters of the receiver's name.
- Select the receiver from the drop down.
- Click on the "Text Input Field" and type in your message.
- Click the Camera icon at the left side of the composed message.
- From Photos, go to the right folder.
- Select the picture you want to send.
- Click Choose and then send.

How to Hide Alerts in Message app on your iPhone XR

- Go to the **Message app** on your iPhone.
- Open the conversation you wish to hide the alert.

- Click on the (i) button at the upper right corner of the page.
- Among the options, one of it is **'Hide alerts'**, move the switch to the right to turn on the option (the switch becomes green).
- Select **'Done'** at the right upper corner of your screen. You are good to go!

How to Set up Your Device for POP3 Email

- From Settings, go to **Accounts and Password.**
- Click on **Add account.**
- Select your service provider from the list or click on others If your service provider is not on the list.
- Select **Add Mail Account.**
- Input your details, name, email address and password.
- Under Description, put in your desired name.
- Click Next at the top right corner of the page.
- The next screen is a confirmation that your email has been set up.
- Follow the on-screen instructions to enter in any extra information.

How to Set up Your Device for IMAP Email

- From Settings, go to **Accounts and Password.**
- Click on **Add account**.
- Select your service provider from the list or click on others If your service provider is not on the list.
- Select **Add Mail Account.**
- Input your details, name, email address and password.
- Under Description, put in your desired name.
- Click Next at the top right corner of the page.
- The next screen is a confirmation that your email has been set up.
- Follow the on-screen instructions to enter in any extra information.
- After this, select **IMAP,**
- Under host name, type in the name of your email provider's incoming server.
- Fill in the username and password for your account.
- Under outgoing host server, type in the name of your email provider's outgoing server.

- Click **Next.**
- Select **Save** at the top right of the screen to save your email address.

How to Set up Your Device for Exchange Email

- From Settings, go to **Accounts and Password.**
- Click on **Add account.**
- Select **Exchange** as your email service provider.
- Input your email address.
- Under Description, put in your desired name.
- Click on **Sign In.**
- Input your email password on the next screen.
- Click on **Sign In.**
- Move the indicator next to the needed data type to enable or disable data synchronization.
- Select **Save** at the top right of the screen to save your email address.

How to Create Default Email Account

- From Settings, go to **Mail** at the bottom of the page.
- Click on **Default Account.**

- On the next screen, click on the email address you wish to set as default.

How to Delete Email Account

- From Settings, go to **Accounts and Password.**
- Click on the email address you want to delete.
- Select **Delete Account** at the bottom of the page.
- On the next screen, click on **Delete from my iPhone.**

How to Compose and Send Email

- From the Home screen, select the Mail icon.
- Click on the back arrow at the top left of the screen.
- Select the email address you want to send the email from.
- Click the new email icon at the bottom right side of the screen.
- On the To field, input the receiver email address and the subject of the email.
- Write your email content in the body of the email.
- To insert a video or picture, press and hold the text input field until a pop-up menu comes up on the screen.

- Click **Insert Pictures or Videos** from the pop-up and then follow the instructions you see on the screen to attach the media.
- To attach a document, select "**Add Attachment"** and follow the instructions you see on the screen.
- Click on **Send** at the right top of the screen.

CHAPTER 6: Manage Applications and Data

How to Install Apps from App Store

- Open the app store and click on search.
- Type in the name of the app in the search field.
- Click on Search.
- Select the desired app.
- Click on **GET** beside the app and follow the steps on the screen to install the app. For paid apps, click on the price to install.

How to Uninstall an App

To uninstall an app,

- click and hold the app until it begins to shake.
- Click on the **Delete** option, then select **Delete.**

With this method, every settings and data about the app would be deleted from your phone.

How to Delete Apps Without Losing the App Data

- From the **Settings,** go to **General.**
- Click **iPhone Storage.**

- Click on the app you wish to uninstall and click on **Offload App.**
- Select **Offload App** again to complete.

How to Control Offload Unused Apps

You can set your device to uninstall apps that are not used in a long time. The app would be uninstalled without deleting the data from the phone. Follow the steps below:

- From the **Settings,** go to **iTunes and App Store.**
- At the bottom of the screen, beside "**Offload Unused Apps",** move the switch left or right to control it.

How to Control Bluetooth

- From **Settings,** go to **Bluetooth**
- Move the switch beside **Bluetooth** to switch on or switch off Bluetooth.

- To pair with a mobile device, put on the Bluetooth then click on the device you want to pair and follow the steps on the screen to link.

How to Control Automatic App Update

- From the **Settings,** go to **iTunes and App Store.**
- Beside **"Update" option,** move the switch left or right to control it.
- Move to **"Use Mobile Data",** move the switch left or right to enable or disable.

How to Chose Settings for Background Refresh of Apps

- From the **Settings,** go to **General.**
- Click on **Background App Refresh.**
- Then click on **Background App Refresh** again.
- Select **OFF** to disable.
- To refresh the apps using Wi-fi, select **Wi-fi.**
- Select **Wi-fi and Mobile Data** if you want to be able to refresh using mobile data.

- Use the back button to return to the previous screen.
- For each of the apps listed, move the slide either left or right to enable or disable.

How to configure your iPhone XR for manual syncing

- Using either Wi-fi or USB, let your device be connected to a computer.
- Manually open the iTunes app if it doesn't come up automatically.
- Tap on the iPhone icon on the top- left of the iTunes screen. If you have multiple iDevice, rather than seeing the iPhone icon, you would see menu showing all the connected iDevices. Once the devices are displayed, select your current device.
- Tap on the **Apply** button at the bottom right corner of your screen.

- Tap on the Sync button if it doesn't start syncing automatically.

How to Synchronize using iCloud

- Click on your Apple ID under Settings.
- Click on iCloud.
- Scroll down to **iCloud Drive** and move the switch left or right to enable or disable.
- Under iCloud, click on **Photos.**
- Scroll down to **Upload to My Photo Stream** and slide left to right to activate or disable.

How to manually add or remove music and videos to your iPhone XR

To manually manage your music and videos, you would have to copy the video files and music tracks to the iPhone from the iTunes Library. Follow the steps below to do this:

- Connect your device to your computer or Mac.
- Launch the iTunes app.
- Manually move the media to the left side of the window.

- Release the media on top of the iPhone (Under Devices).
- Now you can drag any of the items from the main window to the sidebar to add to your iPhone from iTunes.

How to Scan Documents Straight to Files App

The new inbuilt scanner allows you to scan your documents, save as PDFs and even choose your preferred folder for storage. The upgraded to iOS 13 brought lots of improvement to the Files app but the one is being able to scan a document to PDF and being able to save it automatically in Files. To use this feature, follow the steps below.

- Go to the Files app.
- From any location in the app, pull down a little to display the options for view (View and Sorting style).
- Click on the 3-dot (...) icon at the left side of your screen, there, you would see the option to scan a document, create a new folder or connect to a server.

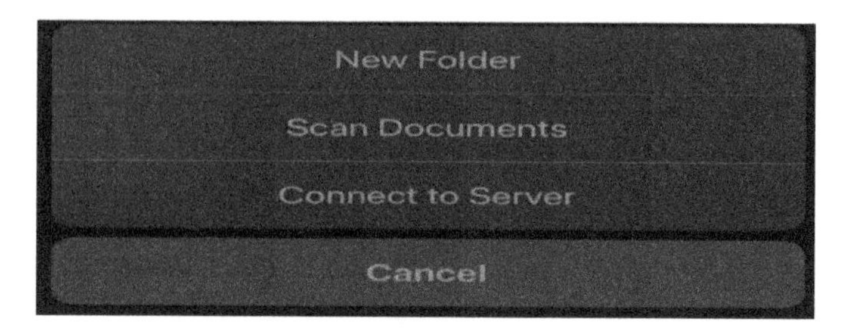

- Scan your receipts or forms as PDFs and get it saved in the cloud folder of your choice.

How to Save Screenshots to the Files App

You can save all your screenshots in your Files app rather than the photos app. You can do this with the steps below:

- Take your screenshot then click on the little preview in the Markup to edit it.
- Then click on **Done.**
- You would see a new option on your screen, added to the **Delete Screenshot** and **Save to Photos** option, you would now have an option to **Save to Files.**
- The **Save to Files** option allows you to save the screenshots in your network folders or iCloud or other Files location outside the Photos app.

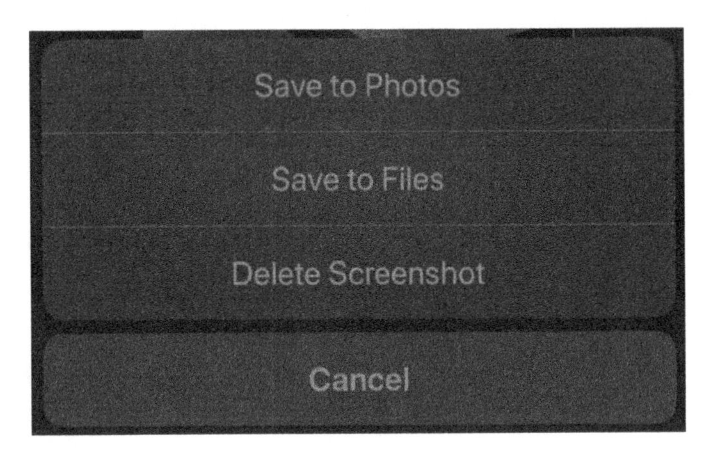

How to Zoom Voice Memos

If you use voice memos a lot, you may have always wanted to have more control over editing or trimming your voice memos. To do this,

- Go to your voice memo
- Click on edit and then you can pinch on the screen to zoom the waveform. This would give you better control and also make it easy for you to scrub through recordings that are very long.

How to Delete Apps in iOS 13

You can now delete apps from your device by

- long pressing on the desired app in your home screen to display an action to rearrange apps.

- Click on this option to get all the apps to wiggle while you would see an **X mark** beside each icon.

- Tap on the **X** beside the apps you wish to remove.

You can also delete apps from the App store. When viewing apps from the app store or updating apps, swipe to the left on any app you want to delete from the list to give you the option to **Delete.** This makes it convenient to delete any app you no longer need from your app store without having to exit the store. Let me explain the steps in details

- Go to the App store app on your device.
- Click on your account picture located at the right top corner of your screen.
- Navigate to the section for apps **Updated Recently.**
- To delete any app that was updated recently. Swipe to the left on the desire app in the list.

- Then click the red button that you see on your screen.

How to Delete Apps from the Update Screen

It is now more convenient to delete apps with the iOS 13. Although the usual way is still active but there is another way to save time.

- Go to the app store. You would see that the updates for app is now in your **Account card.** (click on your picture at the right top of your screen)

- If there are any apps on that list you want to delete, just pull the app to your left and an option would come up to delete the app.

- This applies to all the apps in the list whether it has a pending update or not.

How to Downgrade iOS System on Your iPhone

Did you just recently upgrade your iOS but want to go back to the previous iOS you are familiar with? Here, you will learn tips on how to downgrade without loss of data.

1. First, it is important you back up your device data. I would advise you backup using iCloud as backing up with iTunes can affect your device system when you restore. This would make it easy to restore from iCloud once you have downgraded the iPhone.
   To backup, follow the steps below:

- Go to **Settings>iCloud.**
- Look out for the button titled **'Backup" or "iCloud Backup",** switch it on.
- Ensure your device is connected to Wi-fi and device must be charging during the process.
- Once the backup is done, visit **Settings>Name>iCloud>iCloud Storage>Manage Storage** to confirm the phone backup.

2. Now you are ready to downgrade the system. To downgrade, you need to have a backup file from the iOS you want to switch to. If you don't have any, you can get a standard file downloaded from **Apple Support** to the iOS system. To do this,

- Visit Apple Support, navigate to the Download page
- In the product list, find and select iPhone, then select your desired iOS system.
- Select the option to download to your PC.
- While the file is downloading, update the iPhone on your computer to the most current version.
- Use the lighting cable to connect your phone to your PC.
- On the top left, click the iPhone icon and switch mode to **iPhone Device Panel.**
- Look for the "Restore Backup" button and select it.
- Select the file gotten from the Apple support to downgrade your iPhone to your choice system.
- Once the downgrade is done, you can visit **Setting>General>Software Update** to confirm that the downgrade was done successfully.

3. Once you are done with the downgrade, next is to restore the data saved. Do not worry if you cannot find the data on your iPhone. Just go to the iCloud Backup to restore them. To do this, follow the steps below

- Go to **Settings>General>Reset**.
- Here, you are able to reset the needed data and also "**Erase All Content and Settings**" directly.
- On the App and Data screen, click on "Restore from iCloud"
- Next, input your login details into the iCloud account to choose the backup file you wish to restore.
- Allow it to restore without interruption so that you can get all the files you lost after downgrading.
- That is all there is to downgrading.

CHAPTER 6: Internet and Data

How to Set up your Device for Internet

- Your iPhone is automatically set for internet once the SIM Card is inserted.

How to Use Internet Browser

- Click on the internet browser icon.
- Go to the address bar at the top of the page and input the web address. Then tap Go.
- Click on the menu icon at the bottom of the screen.
- Click on **Add Bookmark.**
- Under Location, click on Favorites and click on Bookmarks.
- Type in the name for the page you want to save and click on save.
- Tap the bookmark icon next to the menu icon.
- Click on the website under bookmark you want to visit.

How to Clear Browser Data

- Go to Settings, then click on **Safari**.

- On the next screen, click on **Clear History and Website Data.**
- From the pop-up, click on **Clear History and Data.**

How to Check Data Usage

- Go to **Mobile Data** under **Settings.**
- Beside **Current Period**, you would see your data usage on the device.
- Under each app, you would see the data usage for those apps.

How to Control Mobile Data

- Go to **Mobile Data** under **Settings.**
- Move the switch beside **Mobile Data** to the right or left to put off or on.
- Scroll to where you have the applications and move the switch beside each app to the right or left to put off or on.

How to Control Data Roaming

- Go to **Mobile Data** under **Settings.**
- Click on **Mobile Data Options.**
- Move the switch beside **Data Roaming** to the right or left to put off or on.

How to Control Wi-fi Setup

- From the top right side of the screen, draw down the screen.
- Click on the Wi-fi icon to enable or disable.
- Move the switch beside **Wi-fi** to the right or left to put off or on.

How to Join a Wi-fi Network

- Go to Settings, then click on Wi-fi.
- Move the switch beside **Wi-fi** to the right to put on the Wi-fi.
- Select your Wi-fi network from the drop down.
- Type in the password and click on **Join.**

How to use your iPhone as a Hotspot

- Go to **Personal Hotspot** under **Settings.**
- Move the switch beside **Personal Hotspot** to the right or left to put off or on.
- IF wi-fi is disabled, click **Turn on Wi-fi and Bluetooth.**
- Select **Wi-fi and USB only** if wi-fi is enabled already.

- Input the wi-fi password beside the field for wi-fi password.
- Select **Done** at the top of the screen.

How to Control Automatic Use of Mobile Data

- Go to **Mobile Data** under **Settings.**
- Move the switch beside **Wi-fi Assist** to the right or left to put off or on.

How to sign into iCloud on your iPhone XR.

- Go to the **Settings app.**
- At the top of your screen, click on **Sign in to your iPhone.**
- Enter your Apple ID email address and password.
- Then click on **Sign In.**
- Next screen would ask for your device passcode if you set up one.
- Set the iCloud Photos the way up like them.
- Switch **Apps using iCloud** on or off, however you want it.

How to Sign Out of iCloud on Your iPhone XR

- From the **Settings app,** click on **Apple ID.**
- Click on **Sign Out** at the bottom of the screen.
- Input your Apple ID password then select **Turn Off.**
- Chose the data you would like to keep a copy of on your iPhone and move the switch on.
- At the top right corner of your screen, click on **Sign Out**.
- Click on **Sign Out** to confirm your decision.

How to Troubleshoot if iCloud isn't Working

If your iCloud isn't working, follow the steps below:

- Ensure the Wi-fi is connected and strong as this is usually the main reason for iCloud backup to not respond.
- Once done, confirm that you have enough space in the cloud. Apple provides only 5G free. If you have used up the spaces, clear the files you don't need or rather back them up with iTunes then remove them from the iCloud backup.
- If you do not wish to delete any information, next step would be to purchase additional room in the iCloud. Please see the pricing below.

  50 GB per month: 0.99 USD

  200GB per month: 2.99 USD

  2 TB per month: 9.99 USD

- Lastly, remove any unwanted data from the iPhone or computer before you perform the iCloud backup.

How to share a calendar on iPhone XR via iCloud

To share your calendar on your iPhone, it is important to first of all turn on the iCloud for calendar option. Kindly follow the steps below:

- On your iPhone, go to **'settings'**
- Click on your device name and select "**iCloud**"
- Then turn on "**Calendars**"

After this is done, you can now share your calendar by following these steps

- Open the "**Calendar**" app on your device.
- At the bottom of your screen, select "**calendars**".
- You would see an "**info**" icon next to the calendar you want to share, click on the icon.

1. Select the **'add person**" option on the screen then pick the people you wish to share the calendar with.
2. Tap "add" followed by "Done" at the top of your screen.

CHAPTER 7: Using Safari

How to Auto Close Open Tabs in Safari on iOS 13

Launch the Safari browser on your iPhone and click on the **View Tabs.** If you are like me, you probably have several open tabs from search results to opened social media posts and so on. Some you want to close but could be quite tiring to close each individually. Thankfully, iOS added a feature to automatically close open tabs in the Safari after a defined time. Follow the steps to activate

- Go to the settings app.
- Navigate to Safari setting and click on it.

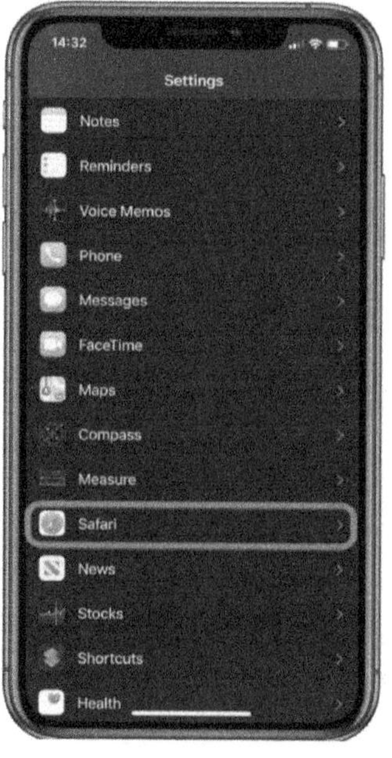

- You would see several options, navigate to the **Tabs** option, then click on **Close Tabs.**
- On the next screen, you would find further options, by default the selection is set to **Manually.**
- You can set to your preferred either **After One Week, After One Day or After One Month.**

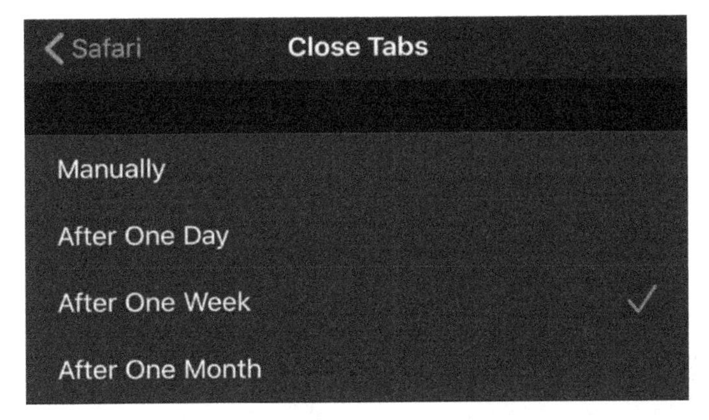

- There is no right or wrong selection. All depends on your usage, for me, I know that most times I never get to go back to the tabs after a week which is why I selected the **After one week** option.

How to Change the Default Location for Downloads from Safari

by default, every download from Safari goes to the **Download folder** in the iCloud drive, but you can modify the location to your preference even if you want it saved in your phone's local storage. To do this,

- Go to settings.
- Click on **Safari.**
- Go to the **General** option and you would see option for **Downloads.**
- Here, you can choose the folders you want the downloaded files to go to.

How to Access Website Settings for Safari

the iOS 13 launch brought some additional features to the Safari browser. One of such change is being able to customize settings for individual sites. Similar to what we have on Safari on Mac, you can now modify different security and viewing options for different websites from the website settings. Safari would them automatically apply the settings so that you do not have to repeat them. I have highlighted the steps below

- Go to a site that you visit regularly.
- Click on the "aA" icon at the left top corner of your screen to show the **View menu** of the website.
- Then click on **Website settings.**

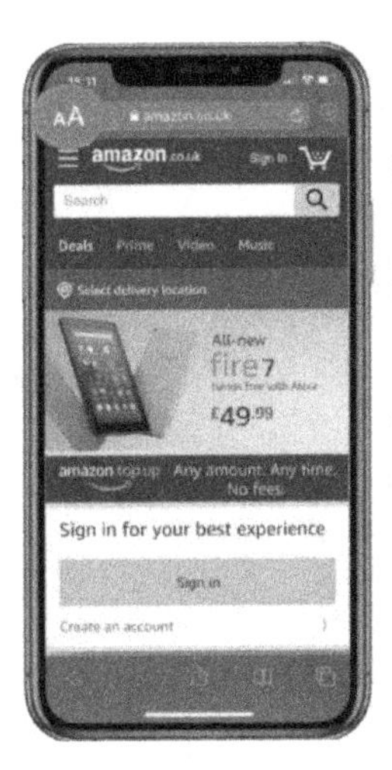

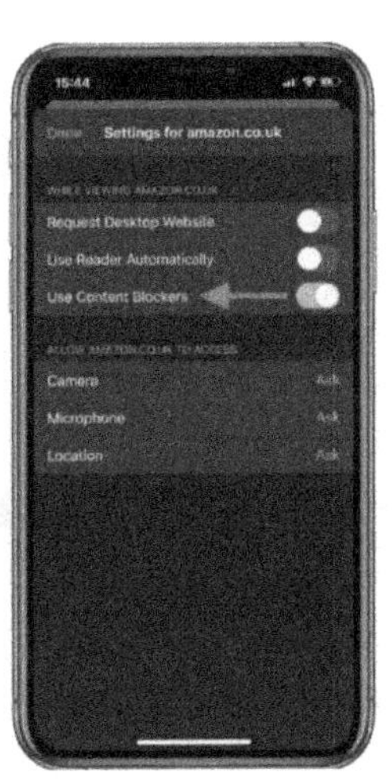

- **Reader Mode** option helps to make online articles more readable by removing extraneous web page contents from it. You can enable this icon by clicking on **"Use Reader Automatically"** to activate this feature by default.

- **Request Desktop Website:** click on this to view original desktop versions of a website on your mobile device.
- **Camera, Microphone, Location:** these last 3 options allows you to choose if you want sites to have access to your device microphone, camera as well as if the sites should be able to know your location. You can choose either **Deny** or **Allow,** but if you would rather change your options per time, then you should select the option for **Ask.** This way, whenever sites want to access these features, Safari would first seek your consent.

How to Modify when the Downloaded File List in Safari is Cleared

The upgrade introduced a Download manager in the mobile version of the Safari browser similar to what is obtainable in Windows and Mac. By default, the list is cleared at the end of each day, however, you can set it to clear the list once the download is done or go for the manual way of clearing lists.

- Go to the settings app.
- Click on **Safari.**

- Then click on **Downloads.**
- Click on **Remove Download List Items.**
- Select any of the options on your screen, Upon successful download, After one day, or Manually.

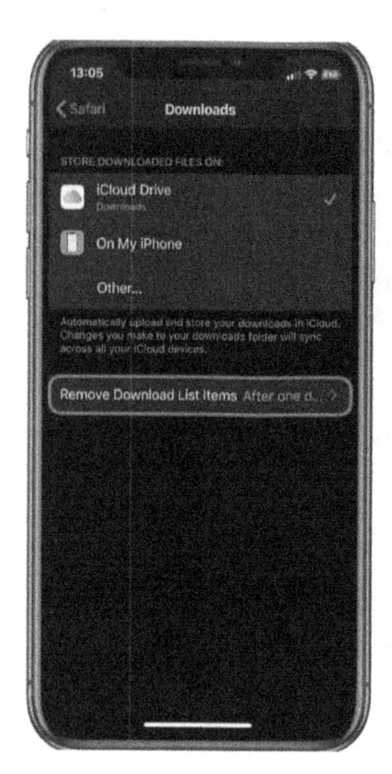

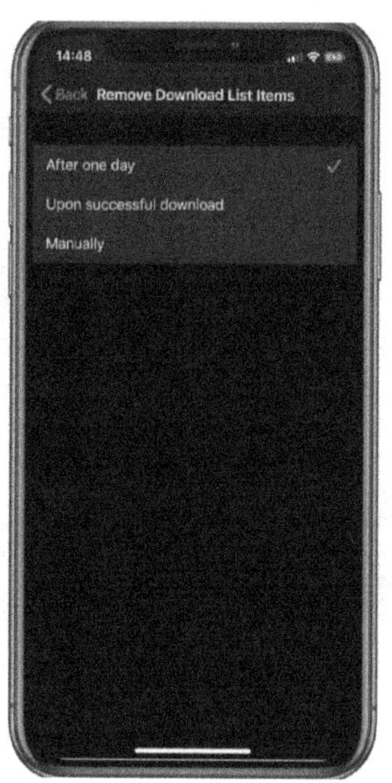

By default, all downloaded files are saved in the **Download folder** of the Files app but you can modify this by selecting an alternative storage location in the settings screen.

How to Access Safari Download Manager

Those who use the Safari desktop version would be more familiar with the Downloads pane in the browser, which informs you on items that have been downloaded and that are currently downloading. Now you can see such with the mobile version of the browser. When you want to download a file, you would see a little download icon at the right top corner of your screen.

Click on the icon to see the status of your downloads, click on the magnifying glass close to the downloaded file to go to the folder where the download is located either on the cloud or on your iPhone.

How to Modify Where Downloaded Files from Safari are Saved

By default, all downloaded files are saved in the **Download folder** of the Files app but you can modify this by selecting an alternative storage location with the steps below:

- Go to the settings app.
- Click on **Safari.**
- Then click on **Downloads.**

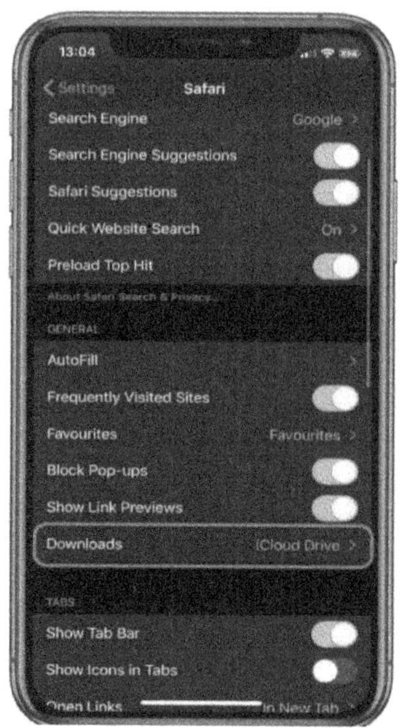

- You can then make your choice from the available options: On My iPhone, iCloud Drive, or in another location that you want.

How to Disable Content Blockers Temporarily in Safari

Content blockers are used to stop ads like banners and popups from loading on any website you visit. It may also disable beacons, cookies and other to protect your privacy and prevent the site from tracking you online. Sometimes, the feature may block an element that you need to access like a web form. If you notice that a useful page element is not coming up because of the content blocker, you can disable it temporarily with the steps below:

- Go to the Safari browser and type in the desired site to visit
- Click on the "aA" icon at the left top corner of your screen to show the View menu of that site.
- Click on" **Turn Off Content Blockers."**

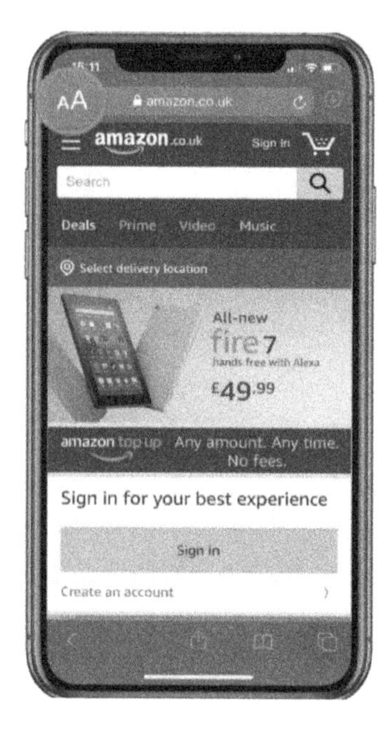

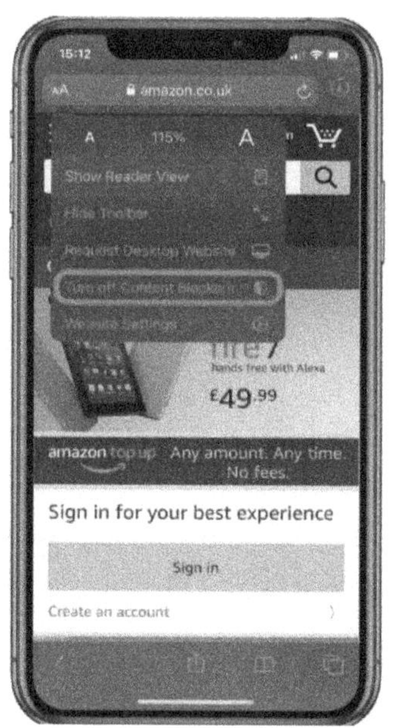

- If you want this disabled for a particular website only, click on **Website Settings** and then move the switch beside **Use Content Blockers** to the left to disable it.

How to Enable Content Blockers in Safari

- Go to the settings app.
- Click on **Safari.**

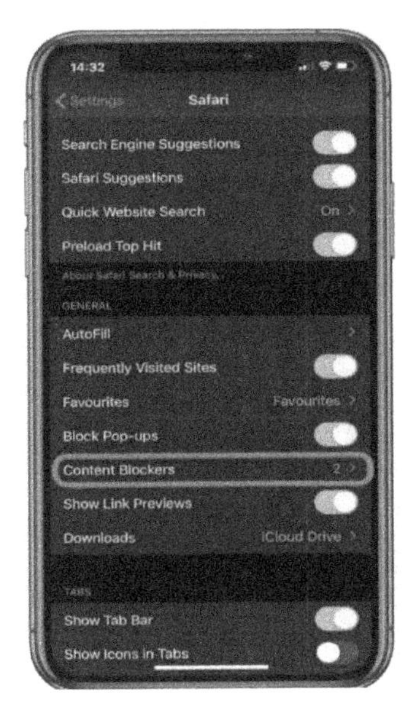

- Go to the **General** option and click on **Content Blockers.**
- Move the switch beside it to the right to enable the option

Note: this option would not be available if you do not install a minimum of one 3rd party content blocker from the store.

How to Share or Save a Safari Web Page as a PDF

this option is only available with the Safari browser and does not include other third-party browsers. Follow the steps to access this feature.

- Open the Safari app on your device.
- Go to the webpage you want to save as PDF.
- Press both the Sleep/wake button and the Home button at same time to take a screenshot.
- If your device does not have a home button, use both the volume up and the power button to take your screenshot.
- You would see a preview of the screenshot at the left lower side of the screen.
- Click on the preview to launch the **Instant Markup Interface,** you have only 5 secs before this screen disappears.
- Click on the **Full-Page** option in the right upper corner of the Markup interface.
- Click on **Done** and then select **Save PDF to Files** to save as PDF.

- Click on the **Share** button to share the PDF and choose who and how you want to share it from that screen.

CHAPTER 8: Conclusion

Now that you have known all there is to know in the iPhone XR, I am confident that you would enjoy operating your device.

The iPhone XR has helped to make things easy and reduce stress only if you have the right knowledge and know how to apply it which I have outlined in this book.

All relevant areas concerning the usage of the iPhone XR from taking out of the box to setup and operations has been carefully outlined and discussed in details to make users more familiar with its operations as well as other information not contained elsewhere.

If you are pleased with the content of this book, don't forget to recommend this book to a friend.

Thank you.

CPSIA information can be obtained
at www.ICGtesting.com
Printed in the USA
LVHW032332230420
654335LV00004B/862